Higher Complex Torsion and the Framing Principle

Memoirs of the American Mathematical Society

Number 835

Higher Complex Torsion and the Framing Principle

Kiyoshi Igusa

September 2005 • Volume 177 • Number 835 (third of 4 numbers) • ISSN 0065-9266

American Mathematical Society
Providence, Rhode Island

2000 *Mathematics Subject Classification.*
Primary 57R45; Secondary 57R50, 19J10.

Library of Congress Cataloging-in-Publication Data

Igusa, Kiyoshi, 1949–
Higher complex torsion and the framing principle / Kiyoshi Igusa.
p. cm. — (Memoirs of the American Mathematical Society, ISSN 0065-9266 ; no. 835)
"Volume 177, number 835 (third of 4 numbers)."
Includes bibliographical references.
ISBN 0-8218-3773-7 (alk. paper)
1. Reidemeister torsion. 2. Differentiable mappings. 3. Diffeomorphisms. I. Title. II. Series.

QA3.A57 no. 835
[QA613.4]
510 s—dc22
[514′.72] 2005048025

Memoirs of the American Mathematical Society

This journal is devoted entirely to research in pure and applied mathematics.

Subscription information. The 2005 subscription begins with volume 173 and consists of six mailings, each containing one or more numbers. Subscription prices for 2005 are \$606 list, \$485 institutional member. A late charge of 10% of the subscription price will be imposed on orders received from nonmembers after January 1 of the subscription year. Subscribers outside the United States and India must pay a postage surcharge of \$31; subscribers in India must pay a postage surcharge of \$43. Expedited delivery to destinations in North America \$35; elsewhere \$130. Each number may be ordered separately; *please specify number* when ordering an individual number. For prices and titles of recently released numbers, see the New Publications sections of the *Notices of the American Mathematical Society.*

Back number information. For back issues see the *AMS Catalog of Publications.*

Subscriptions and orders should be addressed to the American Mathematical Society, P. O. Box 845904, Boston, MA 02284-5904, USA. *All orders must be accompanied by payment.* Other correspondence should be addressed to 201 Charles Street, Providence, RI 02904-2294, USA.

Memoirs of the American Mathematical Society is published bimonthly (each volume consisting usually of more than one number) by the American Mathematical Society at 201 Charles Street, Providence, RI 02904-2294, USA. Periodicals postage paid at Providence, RI. Postmaster: Send address changes to Memoirs, American Mathematical Society, 201 Charles Street, Providence, RI 02904-2294, USA.

This publication is indexed in *Science Citation Index*®, *SciSearch*®, *Research Alert*®,
CompuMath Citation Index®, *Current Contents*®*/Physical, Chemical & Earth Sciences.*
Printed in the United States of America.

∞ The paper used in this book is acid-free and falls within the guidelines
established to ensure permanence and durability.
Visit the AMS home page at `http://www.ams.org/`

10 9 8 7 6 5 4 3 2 1 10 09 08 07 06 05

Contents

Abstract

We prove that higher Franz-Reidemeister (FR) torsion satisfies the transfer property and a formula known as the "Framing Principle" in full generality. We use these properties to compute the higher FR–torsion for all smooth bundles with oriented closed even dimensional manifold fibers. We also show that the higher complex torsion invariants of bundles with closed almost complex fibers are multiples of generalized Miller-Morita-Mumford classes.

Received by the editor June 29, 2004.

2000 *Mathematics Subject Classification.* Primary 57R45; Secondary 57R50, 19J10.

Key words and phrases. higher Franz-Reidemeister torsion, almost complex manifolds, diffeomorphisms, twisting cochains, superconnections, polylogarithms, transfer.

Supported by NSF Grants DMS-0204386, DMS-0309480.

Introduction

This manuscript is a combination of two papers. The first paper is about "higher complex torsion," which is a generalization of higher Franz-Reidemeister (FR) torsion to smooth bundles with almost complex fibers. The rest of the manuscript is a summary of basic facts about higher Franz-Reidemeister torsion. Among these "basic facts" there are several new results, such as the general Framing Principle and its applications and the transfer theorem, but the rest was already explained with full details in [**Igu02**]. A review of the stability theorem is also included at the end.

There are several reasons why these two papers appear together as one manuscript. One is heuristic. The general Framing Principle allows us to compute the higher FR–torsion for smooth bundles with even dimensional fibers. This is needed to show that the complex torsion is a generalization of the "real torsion." The second reason is that the proof of the main theorem about higher complex torsion (the "Complex Framing Principle") is the same as the proof of the Framing Principle. Finally, the facts about higher FR–torsion are not widely known, so it would be difficult for anyone to read the article about higher complex torsion without a summary of the definition and main properties of the higher FR–torsion.

0.1. Higher Franz-Reidemeister torsion

Higher Franz-Reidemeister torsion is an invariant of smooth bundles

$$M \to E \to B \tag{0.1}$$

where M, E, B are compact smooth manifolds. In this introduction we will concentrate on the case of untwisted coefficients and assume that $\pi_1 B$ acts trivially on the rational homology of M. In other words, our bundle (0.1) is classified by a mapping

$$\xi_E : B \to BDiff_0(M)$$

of B into the classifying space of the group $Diff_0(M)$ of diffeomorphisms of M which induce the identity on the rational homology of M.

Here are two examples.

(1) $Diff_0(S^n)$ is the group of orientation preserving diffeomorphisms of S^n.
(2) $Diff_0(\Sigma_g) \simeq \pi_0(Diff_0(\Sigma_g)) = T_g$ is the *Torelli group* if Σ_g is a closed oriented surface of genus g.

Using Morse theory and Waldhausen K-theory, John Klein [**Kle89**] and I [**IK93**], [**Igu02**] constructed a cohomology invariant

$$\tau_{2k}(E) \in H^{4k}(B; \mathbb{R})$$

which is the pullbacks of a universal higher FR–torsion invariant, i.e., $\tau_{2k}(E) = \xi_E^*(\tau_{2k})$ where

$$\tau_{2k} \in H^{4k}(BDiff_0(M); \mathbb{R}).$$

In 1995 Bismut and Lott [**BL95**] constructed similar classes called *higher analytic torsion* classes. Sebastian Goette [**Goe**] explained why these cohomology classes agree in many cases. Basically, it is because both definitions use a canonical A_∞ functor associated to a smooth fiber bundle as we explain later in this introduction.

The homotopy type of the diffeomorphism groups of spheres was determined by Farrell and Hsiang in the 1970's. Using the higher FR–torsion their result can be stated as follows.

THEOREM 0.1 (Farrell-Hsiang[**FH78**]). *(a) For even dimensional spheres we have*

$$H^*(BDiff_0(S^{2N}); \mathbb{R}) \cong \mathbb{R}[\tau_2, \tau_4, \cdots] \cong \mathbb{R}[p_1, p_2, \cdots]$$

in the stable range (through degree $\left\lfloor \min\left(\frac{2N-7}{2}, \frac{2N-2}{3}\right)\right\rfloor$ *by Theorem 7.14).*

(b) For odd dimensional spheres we have twice as much cohomology:

$$H^*(BDiff_0(S^{2N+1}); \mathbb{R}) \cong \mathbb{R}[p_1, \tau_2, p_2, \tau_4, \cdots]$$

in the stable range.

REMARK 0.2 (Newton polynomials). For even dimensional spheres, the relationship between the higher torsion classes τ_{2k} and the Pontrjagin classes p_j is given by

$$\tau_{2k} = (-1)^k \zeta(2k+1) \left(\frac{1}{(2k)!} N_k(p_1, p_2, \cdots) \right)$$

where N_k is the Newton polynomial which expresses $\sum_i x_i^k$ in terms of the elementary symmetric functions of x_i and $\zeta(s) = \sum_{n\geq 1} n^{-s}$ is the Riemann zeta function. For odd spheres a similar relation holds for the restriction of the classes τ_{2k} and p_j to $BSO(2N+2)$. (See Remark 4.2.)

The higher torsion of the Torelli group was first studied by John Klein in [**Kle93**]. Klein conjectured that the higher FR-forsion classes of T_g were proportional to the Miller-Morita-Mumford classes. This was verified by R. Hain, R. Penner and the author in 1992 but the original argument was never published. In [**Igu02**], using the Framing Principle, a precise formula was obtained:

$$\tau_{2k}(T_g^s) = (-1)^k \zeta(2k+1) \frac{\kappa_{2k}}{2(2k)!} \in H^{4k}(T_g^s; \mathbb{R}). \tag{0.2}$$

Here T_g^s is the Torelli group of a surface Σ_g^s of genus g with $s \geq 1$ marked points.

One of the purposes of this paper is to extend this formula to odd Miller-Morita-Mumford classes.

0.2. Construction of τ_k

The basic construction behind the definition of higher Franz-Reidemeister torsion goes back to early work of Ed Brown [**Bro59**] where he defined the concept of a *twisting cochain* ϕ. These can be written as sums of cochains $\phi = \sum_{p\geq 0} \phi_p$ defined on the base space B of a fiber bundle $E \to B$ so that

$$\delta\phi = \phi' \cup \phi$$

where $\phi' = \sum(-1)^p\phi_p$. Here ϕ_p is a p–cochain with coefficients in the degree $p-1$ part of a graded endomorphism ring $\mathrm{End}_R(PR)$ where P is a partially ordered graded set and $PR = R^P$ denotes the free right R–module generated by P. If we add the identity endomorphism of the free R–module PR to ϕ_1 we get an A_∞–functor given on morphisms by $\Phi_p = \phi_p$ for $p \neq 1$, $\Phi_1 = I + \phi_1$ and on objects by $\Phi(X) = (PR, \phi_0(X))$. This is a contravariant A_∞ functor from the category of generic small simplices in B to the differential graded category of finitely generated free R–complexes. (Remark 2.9.)

Bismut and Lott use flat $\mathbb{Z}$–graded *superconnections* of total degree 1. These are linear operators which in local coordinates can be written as $D = d + A$ where $A = \sum_{p\geq 0} A_p$ and A_p is a p–form on B with coefficients in the degree $1-p$ part of a graded endomorphism ring $\mathrm{End}(E)$ where E is a graded vector bundle over B. The superconnection D is *flat* if it satisfies the condition

$$-dA = A \wedge A.$$

In other words, A is an infinitesimal twisting cochain up to sign. An A_∞–functor Φ is given on objects by $\Phi(\sigma) = (E_b, A_0(b))$ where b is the barycenter of $\sigma \subseteq B$. To obtain Φ_p (up to sign) on the morphisms we need to integrate A_p for $p \neq 1$ and $I - A_1$ for $p = 1$. This process is independent of the choice of coordinates. For example, Φ_1 is parallel transport by the connection $d + A_1$ along the edges of the first barycentric subdivision of B. (See [**Goe**], [**Igu**] for more details.)

Brown showed that the homology of the total space of a fiber bundle $E \to B$ is given by the homology of the *twisted tensor product*

$$C_*(B) \otimes_\phi C_*(F)$$

which is the ordinary tensor product complex with boundary map ∂_ϕ which is twisted by a twisting cochain ϕ. If we take coefficients in $\mathbb{Q}$ (or any other field) we can replace $C_*(F)$ by the graded homology $H_*(F;\mathbb{Q})$ which has an A_∞ functor Ψ induced by $\Phi = I + \phi$. If $\pi_1 B$ acts trivially on $H_*(F;\mathbb{Q})$, there is an induced twisting cochain ψ so that $\Psi = I + \psi$ where I is the identity map on $H_*(F;\mathbb{Q})$. Then we get another twisted tensor product

$$C_*(B;\mathbb{Q}) \otimes_\psi C_*(F;\mathbb{Q}) \simeq C_*(E;\mathbb{Q}). \tag{0.3}$$

In the case when $E \to B$ is a smooth bundle with compact manifold fiber M, we can find a fiberwise "generalized Morse function" $f : E \to \mathbb{R}$ which gives a finitely generated cellular chain complex for the fiber over each generic small transverse simplex in B. This Morse theory construction defines another twisting cochain and another twisted tensor product

$$C_*(B) \otimes_\phi C_*(f) \tag{0.4}$$

which is rationally fiberwise quasi-isomorphic to (0.3). In other words, they are equivalent as A_∞–functors. However, they are not simplicially equivalent as twisting cochains on B. There is a well-defined algebraic K-theory obstruction to deforming one twisting cochain into the other. The higher Franz-Reidemeister torsion is a numerical invariant which depends on this K-theory class.

This numerical invariant is given by first interpolating the twisting cochain so that it becomes an infinitesmal twisting cochain, i.e., a flat $\mathbb{Z}$ graded superconnection and integrating an even degree form derived from the odd degree Kamber-Tondeur form. This process, explained in detail in [**Igu02**], is summarized in section 2.8.

0.3. Framing Principle

The Framing Principle is the main tool used to compute the higher Franz-Reidemeister torsion. The original version of the Framing Principle in [**Igu02**] has a restriction on the birth-death singularities of a family of oriented generalized Morse functions. In this paper we remove this condition and prove the Framing Principle for all families of oriented generalized Morse functions.

The general Framing Principle gives us new insight into the nature of higher Franz-Reidemeister torsion. The higher FR–torsion is a sum of two invariants: one is a *tangential invariant* which depends only on the fiber homotopy type of the vertical tangent bundle and the other is an *"exotic" diffeomorphism invariant* which detects different smooth structures on the same fiberwise tangential homeomorphism type.

The tangential invariant $T_{2k}(E)$ is similar in description to the *Miller-Morita-Mumford classes* for oriented surface bundles. This invariant is equal to the higher FR–torsion for all oriented closed even dimensional fibers and equal to zero for oriented closed odd dimensional fibers. Equivalently, the exotic component is zero in the first case and equal to the entire higher torsion invariant in the second case.

For oriented even dimensional fibers with boundary, the higher FR–torsion is given by the tangential invariant plus half the torsion of the boundary:

$$\tau_{2k}(E) = \underbrace{\frac{1}{2}\tau_{2k}(\partial E)}_{exotic\ component} \quad \underbrace{+\frac{1}{2}(-1)^k\zeta(2k+1)\frac{1}{(2k)!}T_{2k}(E)}_{tangential\ component}. \tag{0.5}$$

For oriented odd dimensional fibers with boundary the tangential invariant comes entirely from the boundary.

These facts can be summarized by saying that the exotic torsion comes entirely from oriented odd dimensional fibers in all cases. (For oriented even dimensional fibers the exotic torsion is merely detecting the exotic smooth structure of the boundary. For unoriented fibers we go to either the oriented cover or the oriented disk bundle, whichever is odd dimensional.)

0.4. Complex torsion

Finally we come the the main point of this paper which is to give a generalization of higher FR–torsion for bundles with almost complex (AC) fibers. One of the main motivations is to extend the Klein-Hain-I-Penner relation (0.2) to all of the Miller-Morita-Mumford classes (not just the even ones).

The complex torsion of a bundle $E \to B$ with AC fiber M is a sequence of cohomology classes $\tau_k^{\mathbb{C}}(E,\zeta)_m$ defined using the equivariant higher FR–torsion of the vertical tangent lens space bundle $S^{2n-1}(E)/Z_m$. It is related to the *generalized Miller-Morita-Mumford classes*

$$T_k(E) := tr_B^E(k!ch_k(T^vE)) \in H^{2k}(B;\mathbb{Z})$$

by the following formula assuming that M is a closed AC manifold

$$\tau_k^{\mathbb{C}}(E,\zeta)_m = \frac{1}{2} m^k L_{k+1}(\zeta)\frac{1}{k!}T_k(E) \tag{0.6}$$

where $L_{k+1}(\zeta)$ is the polylogarithm function

$$L_{k+1}(\zeta) = \mathcal{R}\left(\frac{1}{i^k}\sum_{n=1}^{\infty}\frac{\zeta^n}{n^{k+1}}\right).$$

As a special case we obtain the complex torsion of M_g, the mapping class group of a surface of genus g:

$$\tau_k^{\mathbb{C}}(M_g,\zeta)_m = \frac{1}{2} m^k L_{k+1}(\zeta)\frac{\kappa_k}{k!}. \tag{0.7}$$

This is the generalization of equation (0.2) that we were looking for.

Both of the main theorems of this paper, the general Framing Principle and the computation of higher complex torsion, are examples of the *transfer theorem* for higher FR–torsion. This theorem says that the higher FR–torsion commutes with transfer in the sense that, if $D \to E$ is a bundle with fiber Y and $\mathcal{F}$ is a Hermitian coefficient system on D for which Y is acyclic, then the FR–torsion of D as a bundle over B is the transfer of the FR–torsion of D as a bundle over E:

$$\tau_k(D;\mathcal{F})_B = tr_B^E(\tau_k(D;\mathcal{F})_E). \tag{0.8}$$

The transfer theorem and the explanation of the Framing Principle in terms of the transfer were inspired by the work of Dwyer, Weiss and Williams [**DWW03**]. Analogous formulas for higher analytic torsion were given by Xiaonan Ma [**Ma97**].

We note that the transfer formula also explains the presence of polylogarithms in the formula for higher FR–torsion. (See section 3.7.)

I would like to thank Bernhard Keller who directed me to the work of Kadeishvilli [**Kad80**] and thereby also alerted me to the earlier work of my colleague Ed Brown [**Bro59**]. I should also thank Sebastian Goette for explaining his work on higher analytic torsion [**BG01**], [**Goe**]. It was in correspondences with Goette that I realized the importance of proving the Framing Principle in full generality. Finally, I would also like to thank Ed Brown and Gordana Todorov who helped me to sort out my ideas about twisting cochains and A_∞–functors.

Kiyoshi Igusa

CHAPTER 1

Complex torsion

In this chapter we consider smooth bundles

$$M^{2n} \to E \to B \tag{1.1}$$

where the fibers are almost complex (AC) manifolds, i.e., the vertical tangent bundle $T^v E$ has an almost complex structure J. We call this a *bundle with almost complex fibers.* We assume that B is a connected closed manifold but M is only assumed to be compact. We consider separately the cases when M is closed and when it has a boundary.

We will define "higher complex torsion" invariants

$$\tau_k^{\mathbb{C}}(E,\zeta)_m \in H^{2k}(B;\mathbb{R})$$

having properties analogous to the higher FR–torsion invariants for smooth bundles with even dimensional manifold fibers. The definition is a complexified version of Theorems 6.15 and 6.16.

1.1. Definition for closed AC fibers

Although the almost complex fibers of $p : E \to B$ are diffeomorphic they do not necessarily have corresponding almost complex structures (i.e., the diffeomorphism does not preserve J). Thus bundles (1.1) are not classified by the group of diffeomorphisms of M which preserve its almost complex structure. We use instead a standard simplicial construction for the classifying space.

DEFINITION 1.1 (the classifying space $\mathcal{M}_{\mathbb{C}}(M)$). If M^{2n} is an even dimensional real manifold let $\mathcal{M}_{\mathbb{C}}(M)$ be the geometric realization of the simplicial set whose k–simplices are manifolds E diffeomorphic to $M \times \Delta^k$ together with an almost complex structure on the vertical tangent bundle of E over Δ^k.

There is a canonical bundle

$$M \to E_{\mathbb{C}}(M) \to \mathcal{M}_{\mathbb{C}}(M) \tag{1.2}$$

which has an almost complex structure on its vertical tangent bundle.

PROPOSITION 1.2 ($\mathcal{M}_{\mathbb{C}}(M)$ as classifying space). *Any smooth bundle $M \to E \to B$ with almost complex fibers is classified by a map $B \to \mathcal{M}_{\mathbb{C}}(M)$ which is unique up to homotopy.*

PROOF. Any smooth triangulation of B gives a map $B \to \mathcal{M}_{\mathbb{C}}(M)$. Any smooth triangulation of $B\times\partial I$ extends to a triangulation of $B\times I$ giving a homotopy between any two classifying maps. □

Since all complex bundle admit Hermitian metrics which are unique up to isotopy we can construct the vertical tangent sphere bundles:

$$S^{2n-1} \to S^{2n-1}(E) \to E.$$

The circle group $S^1 = U(1)$ acts fiberwise on the sphere bundle $S^{2n-1}(E)$ so we get a free action of every cyclic group

$$Z_m = \{\zeta \in \mathbb{C} \mid \zeta^m = 1\} \subseteq S^1 = U(1).$$

The quotient is a lens space bundle over E:

$$S^{2n-1}/Z_m \to S^{2n-1}(E)/Z_m \to E.$$

Composing with the projection $E \to B$ we get a bundle over B:

$$S^{2n-1}(M)/Z_m \to S^{2n-1}(E)/Z_m \to B. \tag{1.3}$$

The fiber $S^{2n-1}(M)/Z_m$ is a lens space bundle over M.

For any $\zeta \in Z_m$ let ρ_ζ be the one dimensional representation of $\pi_1(S^{2n-1}(M)/Z_m)$ given by composing the canonical epimorphism

$$\pi_1(S^{2n-1}(M)/Z_m) \twoheadrightarrow Z_m$$

with the homomorphism $Z_m \to Z_m$ which sends the generator $e^{2\pi i/m}$ to ζ. When $\zeta \neq 0$ the lens space S^{2n-1}/Z_m has trivial homology with coefficients ρ_ζ. By the spectral sequence of the fibration $S^{2n-1}/Z_m \to S^{2n-1}(M)/Z_m \to M$ this implies that

$$H_*(S^{2n-1}(M)/Z_m; \rho_\zeta) = 0.$$

Consequently, the higher torsion of (1.3) with coefficients in ρ_z is defined for $z \neq 1$.

DEFINITION 1.3 (higher complex torsion for closed fibers). We define the *higher complex torsion* of a bundle with closed almost complex fibers (1.2) to be the family of cohomology classes $\tau_k^{\mathbb{C}}(E,\zeta)_m \in H^{2k}(B;\mathbb{R})$ for all $k \geq 1$, $m \geq 2$ and nontrivial m–th roots of unity ζ given by

$$\tau_k^{\mathbb{C}}(E,\zeta)_m := -\frac{1}{2}\tau_k(S^{2n-1}(E)/Z_m; \rho_\zeta). \tag{1.4}$$

For the universal bundle (1.2) for a closed $2n$–manifold M this gives the *universal higher complex torsion classes*

$$\tau_k^{\mathbb{C}}(\zeta)_m \in H^{2k}(\mathcal{M}_{\mathbb{C}}(M);\mathbb{R}).$$

If the action of $\pi_1 B$ on the rational homology of M is upper triangular then so is its action on $S^{2n-1}(M)/Z_m$ by the Gysin sequence:

$$\cdots \to H_{p+1}(M;\mathbb{Q}) \to H_{p+2n}(S^{2n-1}(M)/Z_m;\mathbb{Q}) \to H_{p+2n}(M;\mathbb{Q}) \to \cdots.$$

So, $\tau_k(E,1)_m$ can be defined by (1.4) in that case.

These invariants are related by the following transfer formula which follows from the transfer formula for cyclic groups (3.31).

PROPOSITION 1.4 (transfer formula for complex torsion). *If z is a nontrivial m–th root of unity then*

$$\tau_k^{\mathbb{C}}(E,z)_m = \sum_{\zeta^r = z} \tau_k^{\mathbb{C}}(E,\zeta)_{mr}. \tag{1.5}$$

This also holds for $z = 1$ if $H_*(M;\mathbb{Q})$ is $\pi_1 B$–upper triangular.

The formula for higher complex torsion (1.4) is a complexified version of Theorem 6.15 where, instead of taking the normal bundle, we take the tangent bundle and changed the sign of the torsion. The following theorem shows that higher complex torsion is a generalization of higher FR–torsion.

THEOREM 1.5 (complex torsion generalizes real torsion). *Suppose that the action of $\pi_1 B$ on $H_*(M;\mathbb{Q})$ is upper triangular. Then the right hand side of (1.4) is defined for $m = 1$ and equal to $\tau_{2k}(E)$ for all $k \geq 1$. We write:*

$$\tau_{2k}^{\mathbb{C}}(E,1)_1 = \tau_{2k}(E).$$

PROOF. This follows immediately from Theorem 6.15 and the following lemma. □

LEMMA 1.6 (torsion of complementary sphere bundles). *Let $S(E)$ be the vertical normal sphere bundle of E (as in Theorem 6.15) and suppose that the action of $\pi_1 B$ on $H_*(M;\mathbb{Q})$ is upper triangular. Then*

$$\tau_{2k}(S(E)) + \tau_{2k}(S^{2n-1}(E)) = 0.$$

REMARK 1.7 (true definition of higher complex torsion). It would not be necessary to invoke this lemma if we used the sphere bundle of the complex vertical normal bundle of E instead of that of the vertical tangent bundle as in Definition 1.3. In other words, it makes more sense to use the following equivalent definition.

$$\tau_k^{\mathbb{C}}(E,\zeta)_m = \frac{1}{2}\tau_k(S(E)/Z_m;\rho_\zeta).$$

PROOF. The Whitney sum of the vertical tangent bundle and the vertical normal bundle is trivial of real dimension $2N$. This implies that the trivial sphere bundle $E \times S^{2N-1}$ over E is the fiberwise join of the tangent sphere bundle $S^{2n-1}(E)$ and the normal sphere bundle $S(E)$. Consequently, we can take any smooth functions

$$f : S^{2n-1}(E) \to [0,1], \quad g : S(E) \to [2,3]$$

and extend them to a smooth function

$$h : E \times S^{2N-1} \to [0,3]$$

which looks locally like $f(x) + t^2$ near $S^{2n-1}(E)$ and $g(y) - t^2$ near $S(E)$ (t being the join parameter) and is nonsingular everywhere else. This makes the cellular chain complex $C_*(h)$ of h into an extension of $C_*(f)$ by $\Sigma^{2n}C_*(g)$. Then by the Splitting Lemma, we have

$$\tau_{2k}(S(E)) + \tau_{2k}(S^{2n-1}(E)) = \tau_{2k}(E \times S^{2N-1}),$$

where

$$\tau_{2k}(E \times S^{2N-1}) = \chi(S^{2N-1})\tau_{2k}(E) = 0.$$

by the product formula (Lemma 3.28). □

1.2. Generalized Miller-Morita-Mumford classes

To state the theorem which computes the higher complex torsion for closed fibers we need the complex version of Definition 6.3. This definition arose out of a conversation with S. Morita.

DEFINITION 1.8 (generalized MMM classes). Given a smooth bundle $p : E \to B$ with almost complex fibers let

$$T_k(E) = tr_B^E\,(k!\,ch_k(T^vE)) \in H^{2k}(B;\mathbb{Z}).$$

PROPOSITION 1.9 (T_k generalized MMM classes). a) When k is even, Definition 1.8 agrees with Definition 6.3.

b) When M is a closed oriented surface, $T_k(E)$ is equal to the Miller-Morita-Mumford class

$$T_k(E) = \kappa_k(E).$$

c) When M is an oriented surface with boundary, $T_k(E)$ is equal to the punctured Miller-Morita-Mumford class

$$T_k(E) = \kappa_k^{\partial}(E).$$

PROOF. By equation (6.3). □

Because of Proposition 1.9 (b) we call $T_k(E)$ the *generalized Miller-Morita-Mumford classes* of E. Note that these invariants are high dimensional stable invariants.

PROPOSITION 1.10 (stability of T_k).

$$T_k(E) = T_k(E \times D^{2N}).$$

Using these generalized MMM classes T_k we have the following formula for the higher complex torsion in the case of closed AC fibers.

THEOREM 1.11 (complex torsion for closed AC fibers). *Suppose that $M^{2n} \to E \to B$ is a smooth bundle with closed almost complex fibers. Then*

$$\tau_k^{\mathbb{C}}(E,\zeta)_m = \frac{1}{2}m^k L_{k+1}(\zeta)\frac{1}{k!}T_k(E)$$

where $L_{k+1}(\zeta)$ is the polylogarithm function

$$L_{k+1}(\zeta) = \mathcal{R}\left(\frac{1}{i^k}\sum_{n=1}^{\infty}\frac{\zeta^n}{n^{k+1}}\right).$$

REMARK 1.12 (polylogarithm term is forced). As we explain in section 3.7, the polylogarithm term is forced by the transfer formula (1.5).

An important special case is the higher complex torsion of the *mapping class group* M_g (the group of isotopy classes of orientation preserving self-diffeomorphisms of a surface Σ_g of genus g).

COROLLARY 1.13 (MMM class as complex torsion). The higher complex torsion of the canonical surface bundle over the classifying space BM_g of the mapping class group M_g is

$$\tau_k^{\mathbb{C}}(M_g,\zeta)_m = \frac{1}{2}m^k L_{k+1}(\zeta)\frac{\kappa_k}{k!}.$$

PROOF OF THEOREM 1.11. Let

$$D = S^{2n-1}(E)/Z_m.$$

This is a lens space bundle over E. So, by the formula for the higher FR–torsion of a lens space bundle (3.33) we have

$$\tau_k(D,\rho_\zeta)_E = -m^k L_{k+1}(\zeta) ch_k(T^vE).$$

By the transfer theorem (5.1), this is related to the torsion of D over B by

$$\tau_k(D;\rho_\zeta)_B = tr_B^E(\tau_k(D;\rho_\zeta)_E) = -m^k L_{k+1}(\zeta)\frac{1}{k!}T_k(E).$$

Multiply this by $-\frac{1}{2}$ to get $\tau_k^{\mathbb{C}}(E;\zeta)_m$. □

1.3. Complex Framing Principle

In order to be able to compute the higher complex torsion we need to rephrase Theorem 1.11 in terms of a smooth function $f : E \to \mathbb{R}$ as in the Framing Principle.

COROLLARY 1.14 (Complex Framing Principle). Let $E \to B$ be a smooth bundle with closed almost complex fibers, let $f : E \to \mathbb{R}$ be any generic smooth function and let $\Sigma(f)$ be the singular set of f. Then

$$\tau_k^{\mathbb{C}}(E,\zeta)_m = \frac{1}{2}m^k L_{k+1}(\zeta)p_*^\Sigma(ch_k(T^vE)|\Sigma(f))$$

where the push-down operator

$$p_* : H^{2k}(\Sigma(f)) \to H^{2k}(B),$$

with real or integral coefficients, is given in section 4.3.

PROOF. This is equivalent to Theorem 1.11 by the commuting diagram in Proposition 4.8. □

We give an example of how the Complex Framing Principle is used and how it compares with the (real) Framing Principle.

Take a complex m–plane bundle ξ over B. Let

$$\mathbb{C}P^{m-1} \to P(\xi) \to B$$

be the associated projective bundle. Since $U(m)$ is connected, $\pi_1 B$ acts trivially on the homology of the fiber. Consequently, all of the higher real and complex torsion invariants are defined.

We first compute the real torsion. By the splitting principle, we may assume that ξ is a sum of line bundles $\xi = \oplus\lambda_i$. Then we choose a Morse function g on $\mathbb{C}P^{m-1}$ with exactly m critical points at the coordinate axes of $\mathbb{C}^m$ so that g is $U(1)^{m-1}$ invariant. This gives a fiberwise Morse function f on $P(\xi)$ with m critical points $x_1,\cdots,x_m$ of indices $0,2,4,\cdots,2m-2$ with corresponding negative eigenspace bundles

$$\gamma_i \cong \mathrm{Hom}_{\mathbb{C}}(\lambda_i,\lambda_1\oplus\cdots\oplus\lambda_{i-1}) \cong \bigoplus_{j<i}\mathrm{Hom}(\lambda_i,\lambda_j).$$

Since there can be no algebraic incidences between these critical points, the associated twisting cochain is zero so $C_*(f)$ has zero torsion. Thus the higher FR–torsion of $P(\xi)$ is given by the correction term in the Framing Principle (4.11):

$$\begin{aligned}\tau_{2k}(P(\xi)) &= (-1)^k\zeta(2k+1)\sum_{i>j} ch_{2k}(\mathrm{Hom}_{\mathbb{C}}(\lambda_i,\lambda_j))\\ &= (-1)^k\zeta(2k+1)\frac{1}{2}ch_{2k}(\mathrm{Hom}_{\mathbb{C}}(\xi,\xi))\end{aligned}$$

Now we compute the higher complex torsion using the Complex Framing Principle 1.14 which tells us to use the entire complex vertical tangent bundle along x_i:

$$\mathrm{Hom}_{\mathbb{C}}(\lambda_{i+1}, \lambda_1 \oplus \cdots \oplus \widehat{\lambda_{i+1}} \oplus \cdots \oplus \lambda_n).$$

Since $\mathrm{Hom}_{\mathbb{C}}(\lambda_{i+1}, \lambda_{i+1})$ is trivial, this is stably equivalent to $\mathrm{Hom}_{\mathbb{C}}(\lambda_{i+1}, \xi)$. Taking the sum over all i we get

$$\tau_k^{\mathbb{C}}(P(\xi), \zeta)_m = \frac{1}{2} m^k \mathcal{R}\left(\frac{1}{i^k}\mathcal{L}_{k+1}(\zeta)\right) ch_k(\mathrm{End}_{\mathbb{C}}(\xi)).$$

We see in this example that the factor of $\frac{1}{2}$ compensates for the fact that we are using the entire vertical tangent bundle instead of just the negative eigenspace bundle.

1.4. Nonempty boundary case

We now assume that $E \to B$ is a smooth bundle with compact almost complex fibers M^{2n} having real dimension $2n$ and nonempty boundary. We assume B is closed so that ∂E is a bundle over B with fiber ∂M.

When the fibers M have nonempty boundary, a correction term should be added to the definition of higher complex torsion.

DEFINITION 1.15 (complex torsion with boundary). We define the *complex torsion* of $E \to B$ to be the family of cohomology classes $\tau_k^{\mathbb{C}}(E; \zeta)_m \in H^{2k}(B; \mathbb{R})$ for all $k \geq 1$, $m \geq 2$ and nontrivial m–th roots of unity ζ given by

$$\tau_k^{\mathbb{C}}(E; \zeta)_m := \frac{1}{2m}\tau_k(\partial E) - \frac{1}{2}\tau_k(S^{2n-1}(E)/Z_m; \rho_\zeta). \tag{1.6}$$

This differs from the closed fiber case by the *boundary correction term*:

$$\frac{1}{2m}\tau_k(\partial E). \tag{1.7}$$

This correction term is zero for k odd and, for k even, is defined only under the assumption that $H_*(\partial M; \mathbb{Q})$ is $\pi_1 B$–upper triangular.

As in the closed fiber case these invariants are related by the following transfer equation whenever the terms are defined.

$$\tau_k^{\mathbb{C}}(E, z)_m = \sum_{\zeta^r = z} \tau_k^{\mathbb{C}}(E, \zeta)_{mr}.$$

The boundary correction term comes from Theorem 6.16 and its proof. Without this correction term, the following theorem would not be true.

THEOREM 1.16 (complex vs real torsion in boundary case). *Suppose that the action of $\pi_1 B$ on $H_*(M; \mathbb{Q})$ and $H_*(\partial M; \mathbb{Q})$ is upper triangular. Then the right hand side of (1.6) is defined for $m = 1$ and equal to $\tau_{2k}(E)$ for all $k \geq 1$. We write:*

$$\tau_{2k}^{\mathbb{C}}(E, 1)_1 = \tau_{2k}(E).$$

PROOF. In Theorem 6.16 and its proof we show that

$$\tau_{2k}(E) = \frac{1}{2}\tau_{2k}(\partial E) + \frac{1}{2}\tau_{2k}(S(E)) - \frac{1}{2}\tau_{2k}(S(E)|_{\partial E})$$

and we also note that the last term is zero by Theorem 6.6 since $S(E)|_{\partial E}$ has closed even dimensional stably parallelizable fibers. This agrees with the definition of $\tau_{2k}^{\mathbb{C}}(E, 1)_1$ by Corollary 1.6. □

As we remarked in 1.7, a more natural way to write the definition of complex torsion is the following

$$\tau_k^{\mathbb{C}}(E;\zeta)_m = \frac{1}{2m}\tau_k(\partial E) + \frac{1}{2}\tau_k(S(E)/Z_m;\rho_\zeta) \tag{1.8}$$

where $S(E)$ is the sphere bundle of the complex vertical normal bundle of E.

As in the case of the generalized MMM classes, the higher complex torsion is stable when defined.

THEOREM 1.17 (stability of complex torsion in boundary case). *If $H_*(M;\mathbb{Q})$ and $H_*(\partial M;\mathbb{Q})$ are $\pi_1 B$ upper triangular then*

$$\tau_k^{\mathbb{C}}(E,\zeta)_m = \tau_k^{\mathbb{C}}(E\times D^{2N},\zeta)_m.$$

PROOF. By stability of T_k, we only need to show that the boundary correction terms for $\tau_k^{\mathbb{C}}(E,\zeta)_m$ and $\tau_k^{\mathbb{C}}(E\times D^{2N},\zeta)_m$ agree.

When M has no boundary the boundary correction term of $\tau_k^{\mathbb{C}}(E\times D^{2N},\zeta)_m$ is

$$\frac{1}{2m}\tau_k^{\mathbb{C}}(E\times S^{2N-1},\zeta)_m = \frac{1}{2m}\chi(S^{2N-1})\tau_k(E) = 0.$$

When M has a boundary this correction term is given by the gluing formula (3.17) as follows.

$$\begin{aligned}\frac{1}{2m}\tau_k(\partial(E\times D^{2N})) &= \frac{1}{2m}\left(\tau_k(\partial E\times D^{2N}) + \tau_k(E\times S^{2N-1}) - \tau_k(\partial E\times S^{2N-1})\right)\\ &= \frac{1}{2m}\tau_k(\partial E).\end{aligned}$$

The other two terms are zero by the product formula (since $\chi(S^{2N-1})=0$). □

Finally, we note that the Complex Framing Principle still holds with the same proof but it does not compute the complex torsion in the boundary case because of the boundary correction term.

THEOREM 1.18 (Complex Framing Principle in boundary case). *Let $f: E\to\mathbb{R}$ be any generic smooth function whose fiberwise gradient points outward along ∂E. Let $\Sigma(f)$ be the singular set of f. Then*

$$\begin{aligned}\tau_k^{\mathbb{C}}(E,\zeta)_m &= \frac{1}{2}m^k L_{k+1}(\zeta)\frac{1}{k!}T_k(E) + \frac{1}{2m}\tau_k(\partial E)\\ &= \frac{1}{2}m^k L_{k+1}(\zeta)p_*^{\Sigma}(ch_k(T^vE)|\Sigma(f)) + \frac{1}{2m}\tau_k(\partial E).\end{aligned}$$

CHAPTER 2

Definition of higher FR–torsion

In this chapter we review the definition of higher Franz-Reidemeister torsion and, more generally, the cellular chain complex functor associated to any fiberwise oriented generalized Morse function.

Suppose we have a smooth bundle

$$M \to E \to B$$

where B is a closed manifold and M is a compact n–manifold with $\partial M = \partial_0 M \coprod \partial_1 M$. Let $\partial E = \partial_0 E \coprod \partial_1 E$ where $\partial_i E$ is a bundle over B with fiber $\partial_i M$.

Given a fiberwise generalized Morse function

$$f : (E, d_0 E, \partial_1 E) \to (I, 0, 1)$$

and a representation $\rho : \pi_1 E \to G$ where G is a subgroup of the group of units of a ring R we construct a family of cellular chain complexes $C_*(f; R)$, with twisted R–coefficients, giving an (R, G)–expansion functor

$$C_*(f; R) : B \to \mathcal{W}h_\bullet(R, G)$$

where $\mathcal{W}h_\bullet(R, G)$ is the Whitehead category. The expansion functor consists of a poset functor and a *twisting cochain* in the sense of [**Bro59**].

In the special case when $(R, G) = (M_r(\mathbb{C}), U(r))$, we have cohomology classes

$$\tau_k \in H^{2k}(\mathcal{W}h^h_\bullet(M_r(\mathbb{C}), U(r)); \mathbb{R})$$

which we call the *universal Franz-Reidemeister torsion invariants*. Here, $\mathcal{W}h^h_\bullet(-,-)$ is the simplicial full subcategory of $\mathcal{W}h_\bullet(-,-)$ given by acyclic chain complexes. Thus, in the basic construction, we obtain higher FR–torsion invariants for $(E, \partial_0 E)$ in the case when we have a Hermitian coefficient system $\mathcal{F}$ on E with respect to which $(M, \partial_0 M)$ is acyclic. They are given by

$$\tau_k(E, \partial_0 E; \mathcal{F}) = C_*(f; \mathcal{F})^*(\tau_k) \in H^{2k}(B; \mathbb{R}).$$

2.1. Generalized Morse functions

A *generalized Morse function* (GMF) on a smooth n–manifold M is a smooth function

$$f : M \to \mathbb{R}$$

having Morse (nondegenerate) and *birth-death* (b-d) singularities. We recall that, at a Morse singularity, the function f can be written in the form

$$f(x, y) = -\|x\|^2 + \|y\|^2 + C \tag{2.1}$$

with respect to local coordinates $(x, y) \in \mathbb{R}^i \times \mathbb{R}^{n-i}$ for M and C is a constant (the critical value). At a birth-death point, f can be written as

$$f(x, y) = -\|(x_1, \cdots, x_{i-1})\|^2 + x_i^3 + \|y\|^2 + C. \tag{2.2}$$

We choose a Riemannian metric on M which we will assume to be *standard* in a coordinate neighborhood of each critical point. By this we simply mean that the x–plane is perpendicular to the y–plane at each point in a small neighborhood. In the case of a birth-death point we also require the x_i direction be perpendicular to the tangent plane spanned by $\frac{\partial}{\partial x_j}$ for $j = 1, \cdots, i-1$ in a small neighborhood of the b-d point.

To begin, we choose the metric on M so that the coordinate chart given by (x, y) is an isometry. We say in that case that the coordinates are *isometric*. Then the gradient of f will be the transpose of the derivative, i.e,

$$\nabla f(x, y) = (-2x, 2y)^t = -\sum 2x_j \frac{\partial}{\partial x_j} + \sum 2y_k \frac{\partial}{\partial y_k}$$

near a Morse point and

$$\nabla f(x, y) = (-2x_1, \cdots, -2x_{i-1}, 3x_i^3, 2y_1, \cdots, 2y_{n-i})^t$$

near a b-d point.

If the coordinates are not isometric but the metric is still standard then the subsets of the coordinate neighborhood given by $x = 0$ and $y = 0$ will still be totally geodesic in both cases as will the subset given by $x_i = 0$ and the x_i–axis in the b-d case. Consequently, the union of the set of trajectories of ∇f converging to the critical point is unchanged.

The b-d singularity given in (2.2) has *index* $i-1$ since this is the index of the second derivative $D^2 f$ of f. The vectors $\frac{\partial}{\partial x_j}$ span the *negative eigenspace* of $D^2 f$. The negative eigenspace is a subspace of the tangent space of M at any critical point which depends on the choice of metric. However, it is unique up to convex choice. An *orientation* for f is defined to be an orientation for the negative eigenspace of $D^2 f$ at every critical point of f.

In a generic p–parameter family of GMF's birth-death points will occur on a codimension one subset of the parameter space. Each b-d point will be *generically unfolded* in the sense that the family of functions f_t has the form:

$$f_t(x, y) = -\|(x_1, \cdots, x_{i-1})\|^2 + x_i^3 + t_0 x_i + \|y\|^2 + C \tag{2.3}$$

with respect to parameter coordinates $t_0, t_1, \cdots, t_{p-1}$ and t–dependent local coordinates (x, y) for M. The singular set $\Sigma(f)$ is a smooth manifold given in this coordinate patch by

$$t_0 = -3x_i^2$$

(and all other coordinates equal to zero). Choose a t–dependent metric for M so that the coordinates x, y are isometric for each t. Then, for each t, the metric will be standard at the Morse points of f_t. [If we reparametrize the x_i coordinate so that the nondegenerate quadratic singularities that occur when $t_0 < 0$ have the form $\pm x_i^2$ as required by (2.1) then the coordinate x_i will no longer be isometric, but it will still be perpendicular to the other coordinates.]

Given an oriented Morse function

$$f : M \to [0, 1]$$

with the property that the boundary of M is the disjoint union of $\partial_0 M = f^{-1}(0)$ and $\partial_1 M = f^{-1}(1)$ we get a *cellular chain complex* $C_*(f)$ which is a free $\mathbb{Z}[G]$–complex $(G = \pi_1 M)$ with a basis which is well-defined up to permutation and multiplication by elements of G. The homology of the cellular chain complex is the relative homology

$$H_k(C_*(f)) = H_k(M, \partial_0 M; \mathbb{Z}[G]).$$

In a generic family of GMF's, the orientations of negative eigenspaces should be chosen in such a way that the incidence between the cancelling critical points at a birth-death point is positive ($+g$ for some $g \in G$).

DEFINITION 2.1 (fiberwise oriented GMF). A *fiberwise oriented GMF* on a smooth bundle $E \to B$ is defined to be a smooth function

$$f : E \to \mathbb{R}$$

which is a GMF on each fiber M together with a continuous family of orientations for the negative eigenspace of $D^2 f_t$ so that the incidence coefficients of cancelling critical points is positive.

There are two problems with fiberwise oriented GMF's. The first is that they may not exist and the second is that they may not be unique. For the existence we know that a fiberwise oriented GMF exists for a smooth bundle $E \to B$ whose fiber dimension is greater than or equal to its base dimension:

$$\dim M \geq \dim B.$$

There are also special arguments in the case $\dim M \leq 2$. To make the fiber dimension arbitrarily large, we can take the product with a large even dimensional sphere S^{2N}. This multiplies all higher torsion invariants by 2 (the Euler characteristic of S^{2N}) so we can divide by 2 to get the invariant for the original bundle. (This procedure is justified by Lemma 3.28 below.) We could also take the product with any large dimensional disk and round off the corners.

For the uniqueness problem we take a special type of fiberwise oriented GMF called a fiberwise "framed function."

DEFINITION 2.2 (framed function). A *framed function* on a smooth manifold M is a GMF $f : M \to \mathbb{R}$ together with a framing of the negative eigenspace of $D^2 f$ at each critical point. At a birth-death point of index $i-1$ we also choose the i–th framing vector to be

$$v_i = \frac{\partial}{\partial x_i}$$

in standard coordinates (2.2). A *fiberwise framed function* on a smooth bundle $E \to B$ is a smooth function $f : E \to \mathbb{R}$ together with continuous families of tangent vectors $v_1, v_2, \cdots$ giving the restriction $f_t : M_t \to \mathbb{R}$ of f to each fiber the structure of a framed function.

REMARK 2.3 (orientation of a framed function). A fiberwise framed function gives a fiberwise oriented GMF if we take the orientation given by reversing the framing vectors: $(v_i, \cdots, v_1)$.

THEOREM 2.4 (Framed Function Theorem[**Igu87**]). *The space of framed functions on a compact smooth n–manifold M is $n-1$ connected. Consequently, a smooth bundle $E \to B$ admits a fiberwise framed function $f : E \to \mathbb{R}$ if $\dim M \geq \dim B$ and f is unique up to framed homotopy if $\dim M > \dim B$ (i.e., given two fiberwise framed functions f, g there is a fiberwise framed function on the bundle $E \times I \to B \times I$ which restricts to f, g on $E \times 0$ and $E \times 1$).*

REMARK 2.5 (C^1–local framed function theorem). In [**Igu90**] it was shown (under the same dimension restrictions) that the fiberwise framed function f can be chosen to be arbitrarily C^1–close to any given smooth function g. Furthermore, f can be chosen to be equal to g outside of an arbitrarily small neighborhood of the set on which g is not fiberwise framed. In particular $\Sigma(f)$ will lie in an arbitrarily small neighborhood of $\Sigma(g)$.

2.2. Families of chain complexes

Given a fiberwise oriented GMF f on the total space of a smooth bundle $E \to B$ and a *vertical metric* on E (i.e., a metric on the *vertical tangent space* T^vE) which we always assume is standard near each critical point, we get a family of chain complexes parametrized by B. We call it $C_*(f)$.

If f is chosen to be fiberwise framed then $C_*(f)$ will be unique up to homotopy by the Framed Function Theorem (and the uniqueness up to homotopy of the vertical metric) and we use it to define the higher FR–torsion invariants. If f is not framed we can still recover the higher FR–torsion from the Framing Principle. Thus, it is useful to consider $C_*(f)$ even when it is not canonical.

In this section we will describe what we mean by a "family of chain complexes." For the technical details, we refer to [**Igu02**]. To (hopefully) simplify the presentation we reverse the direction of the arrows. Since every category is homotopy equivalent to its opposite category, this does not change anything.

The idea is simple. We take a small k simplex σ in the base B. Assume first that the function f is Morse over every point in σ. Then the singular set forms a disjoint union of sheets

$$\Sigma(f|\sigma) \cong \Sigma(f_b) \times \Delta^k$$

where $b \in \sigma$ is, say, the barycenter. If f has b-d points over σ we delete all components of $\Sigma(f|\sigma)$ which contain b-d points. The remaining set of sheets forms a *poset* (partially ordered set) P with a *standard ordering* given by $x < y$ if the critical value of x is less than the critical value of y over every point in σ:

$$x < y \Leftrightarrow (\forall t \in \sigma) f_t(x) < f_t(y).$$

In this paper we will not use the standard partial ordering on P. We use instead the (adjusted) *minimal partial ordering* which is given as follows. Let $P^+(\sigma)$ be the set of all components of $\Sigma(f|\sigma)$. I.e., $P^+(\sigma)$ is P with non-Morse components added. We assume that σ is transverse to the *bifurcation set* of f (the set of all $t \in B$ for which f_t is not Morse). So, this will be a finite set. We take the transitive relation on $P^+(\sigma)$ generated by the relation $x < y$ if for some $t \in \sigma$ a trajectory of ∇f_t goes from a point in x up to a point in y.

PROPOSITION 2.6 (minimal partial ordering). *If σ is sufficiently small this transitive relation will also be anti-reflexive. A fortiori, it will satisfy the following. If $x < y$ in $P^+(\sigma)$ then*

$$f_t(x_i) < f_t(y_j)$$

for every $t \in \sigma$, $x_i \in x \cap \Sigma(f_t)$ and $y_j \in y \cap \Sigma(f_t)$.

Let $P = P(\sigma)$ be the subset of $P^+(\sigma)$ of Morse components with the induced partial ordering. Then the proposition above tells us that the standard ordering on P is a refinement of the minimal ordering. The minimal ordering depends on the vertical metric which we always assume is standard near the critical points. Later, in section 2.9, we will adjust the partial orderings of both $P(\sigma)$ and $P^+(\sigma)$.

Note that P is *graded* since every element $x \in P$ has a *degree* given by the index of f at x. However, it is useful not to assume any relationship between the partial ordering and the grading.

Over each vertex v_i of σ we obtain, by standard Morse theory, a cellular chain complex $C_*(f_{v_i})$. Recall that the definition of this complex requires that a path be chosen from the base point of E to each critical point of f_{v_i}. Since Δ^k is contractible, this can be done consistently over every point in σ. We also need an orientation of the topological cells associated to each critical point. This choice is given by the orientation of f.

With these choices, $C_*(f_{v_i})$ is a free based chain complex over some ring R. ($R = \mathbb{Z}[G]$ with $G = \pi_1 E$ is the universal case.) The basis set will be the graded poset P for each vertex v_i. However the boundary map will depend on i. We call the boundary map $\phi_0(i)$. Thus

$$C_*(f_{v_i}) = (PR, \phi_0(i))$$

where $PR = R^P$ denotes the free right R–module generated by P. Note that $\phi_0(i)$ is an endomorphism of PR satisfying the following.

(1) $\phi_0(i)$ is homogeneous of degree -1.
(2) $\phi_0(i)(x)$ is an R–linear combination of elements $y < x$ of P.
(3) $\phi_0(i)^2 = 0$

Any endomorphism of PR satisfying condition (2) will be called *strictly upper triangular.* (If the condition is weakened to $y \leq x$ we call it *upper triangular.*)

Over each edge (v_i, v_j) in σ with $i < j$ we have an upper triangular chain isomorphism

$$\Phi_1(i,j) : C_*(f_{v_j}) = (PR, \phi_0(j)) \xrightarrow{\approx} (PR, \phi_0(i)) = C_*(f_{v_i}).$$

This chain isomorphism is "close" to the identity map on PR so we write it as

$$\Phi_1(i,j) = 1 + \phi_1(i,j).$$

Then $\phi_1(i,j)$ will be a strictly upper triangular endomorphism of PR of degree 0 satisfying the equation

$$\phi_0(i)(1 + \phi_1(i,j)) = (1 + \phi_1(i,j))\phi_0(j),$$

which can also be written as

$$\phi_0(i)\phi_1(i,j) - \phi_1(i,j)\phi_0(j) = \phi_0(j) - \phi_0(i). \tag{2.4}$$

Given three vertices v_i, v_j, v_k with $i < j < k$, the two–parameter family of functions f over the corresponding triangular face of σ gives us a strictly upper triangular chain homotopy

$$\phi_2(i,j,k) : \Phi_1(i,j)\Phi_1(j,k) \simeq \Phi_1(i,k).$$

This can be written in the form

$$\begin{aligned}\phi_0(i)\phi_2(i,j,k) - \phi_1(i,j)\phi_1(j,k) + \phi_2(i,j,k)\phi_0(k)& \\ = \phi_1(j,k) - \phi_1(i,k) + \phi_1(i,j)&.\end{aligned} \tag{2.5}$$

DEFINITION 2.7 (Δ^k–family of chain complexes). Let R be any ring. Then by a *Δ^k–family of chain complexes* over R we mean a pair (P, ϕ) where P is a finite nonnegatively graded poset and ϕ is a system of endomorphisms

$$\phi_p(a_0, \cdots, a_p) : PR \to PR$$

for all $0 \leq a_0 < a_1 < \cdots < a_p \leq k$ which are homogeneous of degree $p-1$, which are strictly upper triangular ($\phi_p(a)(x)$ is a linear comb. of $y < x$ in P) and which satisfy the following (where $\phi_{-1}() = 0$).

$$\sum_{i=0}^{p}(-1)^i \phi_i(a_0, \cdots, a_i)\phi_{p-i}(a_i, \cdots, a_p) = \sum_{i=0}^{p}(-1)^i \phi_{p-1}(a_0, \cdots, \widehat{a_i}, \cdots, a_p). \tag{2.6}$$

Let $\Delta^k(R)$ denote the set of all Δ^k–families of R–complexes where P lies in some fixed universe Ω. These form a simplicial set $\Delta^\bullet(R)$ in an obvious way.

Condition (2.6) can be expressed in the following succinct form where $\phi = \sum \phi_p$ is a sum of cocycles on Δ^k with coefficients in the graded ring $\mathrm{End}_R(PR)$ and $\phi' = \sum(-1)^p \phi_p$.

$$\phi' \cup \phi = \delta\phi. \tag{2.7}$$

REMARK 2.8 (twisting cochains). Ed Brown [**Bro59**] was the first to consider sums of cochains ϕ satisfying (2.6), (2.7). He called them *twisting cochains.* To show that these are the same, we can rewrite the conditions as

$$\partial\phi_p = \phi_{p-1}\partial - \sum_{i=1}^{p-1}(-1)^i \phi_i \cup \phi_{p-i} \tag{2.8}$$

where $\partial\phi_p = \phi_0\phi_p + (-1)^p\phi_p\phi_0$. Since the differential ϕ_0 varies from point to point, ϕ is, strictly speaking, a twisting cochain with coefficients in a differential graded (DG) category whose DG Hom sets have the form

$$\mathrm{Hom}(A, B) = \mathrm{Hom}((PR, \phi_0(A)), (PR, \phi_0(B))). \tag{2.9}$$

To express the relationship between ϕ, R, P we say that ϕ is a twisting cochain *over* R which is *subordinate* to P.

REMARK 2.9 (A_∞–functors). Let Φ_p be given by $\Phi_1 = 1 + \phi_1$ and $\Phi_p = \phi_p$ for $p \neq 1$. If ϕ is a twisting cochain on the nerve of a category $\mathcal{C}$ then Φ is an *A_∞–functor* from $\mathcal{C}^{op}$ to the DG category described above. To see this we convert notation as follows.

$$\begin{aligned}\delta\phi - \phi' \cup \phi &= \sum_{i=0}^{p}(-1)^i \phi_{p-1}\partial_i - \phi' \cup \phi \\ &= \sum_{i=1}^{p-1}(-1)^i \Phi_{p-1}\partial_i - \Phi' \cup \Phi\end{aligned}$$

Then note that:

$$\begin{aligned}\Phi_{p-1}\partial_i &= \Phi_{p-1}(1^{i-1}, m_2, 1^{p-i-1})\\ \Phi' \cup \Phi &= \sum_{i=0}^{p}(-1)^i m_2(\Phi_i, \Phi_{p-i})\\ &= \sum_{i=1}^{p-1}(-1)^i m_2(\Phi_i, \Phi_{p-i}) + m_1(\Phi_p),\end{aligned}$$

where $m_2(f,g) = g \circ f$ ($f \circ g$ in $\mathcal{C}^{op}$) and $m_1 = \partial$. Thus (2.7) becomes:

$$\sum_{i=1}^{p-1}(-1)^i \Phi_{p-1}(1^{i-1}, m_2, 1^{p-i-1}) = m_1(\Phi_p) + \sum_{i=1}^{p-1}(-1)^i m_2(\Phi_i, \Phi_{p-i}).$$

For simplicity we follow the twisting cochain formalism. We will vary the poset to obtain what we call "monomial functor" and later we will generalize the construction to an "expansion functor." Just as in the case of A_∞ functors, these are not true functors. However, we will see that they can be reinterpreted as true functors in Remark 2.17 below.

2.3. Monomial functors

Suppose that $f : E \to \mathbb{R}$ is a fiberwise oriented Morse function and R is an untwisted coefficient ring. Then, we can apply the construction above to every small smooth simplex $\sigma : \Delta^p \to B$ to obtain p–cochains ϕ_p on B with coefficients in the strictly upper triangular degree $p-1$ part of the graded ring $\mathrm{End}_R(PR)$ whose sum $\phi = \sum \phi_p$ satisfies (2.7) above where P is a functor from the category $\mathrm{simp}\, B$ of small smooth simplices in B to the category $\mathcal{P}$ of partially ordered graded sets. (Morphisms in $\mathcal{P}$ are degree 0 order preserving bijections $\alpha : P \to Q$, i.e., $x < y \Rightarrow \alpha(x) < \alpha(y)$ and $\deg(x) = \deg \alpha(x)$.)

We recall that the *category of simplices* $\mathrm{simp}\, X$ of any simplicial set X has as objects all pairs $(\sigma, [k])$ where $k \geq 0$ and $\sigma \in X_k$ and morphisms $(\tau, [j]) \to (\sigma, [k])$ are defined to be (nondecreasing) maps $a : [j] \to [k]$ so that $\tau = a^*(\sigma)$.

This construction needs to be generalized to twisted coefficients R (with $G \subseteq R^\times$ containing the image of $\pi_1 E$) as follows.

DEFINITION 2.10 (the category $\mathcal{P}(G)$). Let $\mathcal{P}(G)$ be the category whose objects are finite partially ordered graded sets. A morphism in $\mathcal{P}(G)$ is a right G–equivariant mapping

$$\widetilde{\alpha} : P \times G \to Q \times G$$

so that the induced map on orbits $\alpha : P \to Q$ is a degree 0 order preserving bijection. We say that $\alpha : P \to Q$ is a *G–morphism* to indicate that it is covered by a G–equivariant mapping $\widetilde{\alpha}$.

If $G \subseteq R^\times$, the free right R–module PR is a functor on $\mathcal{P}(G)$. A G–morphism $\alpha : P \to Q$ induces an isomorphism $\alpha_* : PR \to QR$ by

$$\alpha_*\left(\sum x_i r_i\right) = \sum \pi\widetilde{\alpha}(x_i, 1) r_i$$

where $\pi(x, g)r = x(gr)$. The inverse of α_* will be denoted r (for *restriction*). Conjugation by α_* gives a ring isomorphism

$$(\,)_\# : \mathrm{End}_R(PR) \to \mathrm{End}_R(QR)$$

making $\mathrm{End}_R(PR)$ into a functor on $\mathcal{P}(G)$. If f is strictly upper triangular then so is $f_\#$.

If $f : E \to \mathbb{R}$ is a fiberwise oriented Morse function and R is a coefficient ring twisted by a representation $\pi_1 E \to G \subseteq R^\times$ then the singular set of f, with the minimal partial ordering, gives a functor

$$P = \Sigma(f) : \mathrm{simp}\, B \to \mathcal{P}(G)$$

and for each $p \geq 0$ we get a p–cochains ϕ_p on B with coefficients in the strictly upper triangular degree $p-1$ part of $\mathrm{End}_R(PR)$ whose sum $\phi = \sum \phi_p$ satisfies (2.7). The functor P together with the twisting cochain ϕ forms what we call an (R, G)–*monomial functor* (P, ϕ) on B.

DEFINITION 2.11 ((R, G)–monomial functor). An (R, G)–*monomial functor* on a simplicial set X is defined to be a pair (P, ϕ) where P is a functor

$$P : \mathrm{simp}\, X \to \mathcal{P}(G)$$

and $\phi = \sum \phi_p$ where ϕ_p is a p–cochain on X with coefficients in the strictly upper triangular degree $p-1$ part of $\mathrm{End}_R(PR)$ so that $\delta\phi = \phi' \cup \phi$. We say that ϕ is a *twisting cochain subordinate* to the functor P.

To clarify this definition, when we say that ϕ_p is a p–cochain on X we mean that ϕ_p is a function that takes every element $x \in X_p$ to a strictly upper triangular endomorphism $\phi_p(x) \in \mathrm{End}_R(P(x)R)$. When we write $\delta\phi = \phi' \cup \phi$ we mean:

$$\sum_{i=0}^{n} (-1)^i \phi_{n-1}(\partial_i x)_\# = \sum_{p+q=n} (-1)^p \phi_p(f_p(x))_\# \phi_q(b_q(x))_\#$$

in $\mathrm{End}_R(P(x)R)$. In other words, we are inducing the endomorphisms on the faces of x back up to x and the condition is that these form an element

$$(P(x), \phi_\#) \in \Delta^k(R). \tag{2.10}$$

DEFINITION 2.12 (twisted tensor product). Ed Brown's *twisted tensor product* (giving the homology of the total space [**Bro59**]) is called the *total complex* of (P, ϕ) in [**Igu02**]:

$$C_*(X) \otimes_\phi (PR, \phi_0) = E_*(P, \phi) = \bigoplus_{n \geq 0} \bigoplus_{x \in X_n} x \otimes P(x)R \tag{2.11}$$

with boundary ∂_ϕ given by:

$$\partial_\phi(x \otimes y) = \partial x \otimes r(y) - \sum_{p+q=n} (-1)^p f_p(x) \otimes r(\phi_q(b_q(x))_\#(y))$$

where each r represents the appropriate restriction map. For example,

$$\partial x \otimes r(y) = \sum (-1)^i \partial_i x \otimes \partial_i^*(y).$$

2.4. Filtered chain complexes

This section describes the transition from Morse theory to the algebra of twisting cocycles. The basic idea is to take the total singular complex of $E|\sigma$ for each small k–simplex σ together with the filtration given by a fiberwise oriented GMF. This is algebraically equivalent to the cellular chain complex of the cell complex arising from the Morse function of a representative fiber. However, it has the advantage of being canonical. The theory of filtered chain complexes is joint work of

J. Klein and the author. This includes the existence of minimal filtered subcomplexes. Minimal filtered chain complexes are shown to be isomorphic to twisting cochains in [**Igu02**]. However, I was not aware of the connection to Brown's paper [**Bro59**] until after I talked to Bernhard Keller.

Suppose that, as above, we have a fiberwise oriented GMF $f : E \to \mathbb{R}$ and a small smooth k–simplex σ in B which is transverse to the bifurcation set of f. Then we have an open subset of $E|\sigma$ associated to closed subsets of the poset $P^+(\sigma)$ of all components of $\Sigma(f|\sigma)$.

A subset Q of a poset P is called *closed* if it contains every element of P which is less than an element of Q. The complement of Q in P is denoted P/Q. (It is the *quotient.*) For $P = P^+(\sigma)$, this means that there are no trajectories of ∇f_t going up from a point in P/Q up to a point in Q for any $t \in \sigma$.

For every closed subset Q^+ of $P^+(\sigma)$ let $E^{Q^+}|\sigma$ be the union over all $t \in \sigma$ of the set all points x in M_t so that the trajectory of ∇f_t passing through x contains an element of the set

$$\partial_0 M_t \coprod Q^+ \tag{2.12}$$

in its closure. In other words, either x is an element of (2.12) or the trajectory of ∇f_t through x comes from an element of (2.12). Another description of the same set $E^{Q^+}|\sigma$ is that it is the complement in $E|\sigma$ of the set of points multiply incident over $P^+(\sigma)/Q^+$.

Note that in the special case of the empty set $Q^+ = \emptyset$, the space $E^{\emptyset}|\sigma$ is homotopy equivalent to $\partial_0 E|\sigma$.

We are now ready to define the "filtered chain complex" for f over σ. This will be additional structure on the total singular complex of $E|\sigma$ with twisted R–coefficients. The idea is that $E|\sigma$ is homotopy equivalent to a relative cell complex

$$\partial_0 E|\sigma \cup e^{i_1} \cup e^{i_2} \cup \cdots$$

where cells correspond to Morse elements of the poset $P^+(\sigma)$ and pairs of cancelling cells correspond to birth-death elements of $P^+(\sigma)$. The filtration picks out the subcomplexes of this chain complex which correspond to the subcomplexes of the relative chain complex. These subcomplexes are indexed by the closed subsets $Q^+ \subseteq P^+(\sigma)$. If $Q = A \coprod x$ where x is Morse and maximal in Q then the subcomplex for Q is given by attaching one cell to the subcomplex for A. The quotient of the corresponding chain complexes is called a "Morse layer" and is chain isomorphic to $\Sigma^i R$ where i is the index of x. This isomorphism is given by a cohomology class on the Morse layer which we call the "dual class."

As part of the definition of a filtered chain complex we also include all of the above structure for every face τ of σ.

DEFINITION 2.13 (filtered chain complex). The *filtered chain complex* of f over σ is defined to be the total singular complex $S_*(E|\sigma;\rho)$ of $E|\sigma$ with coefficients R twisted by a representation

$$\rho : \pi_1 E \to G \subseteq R^\times$$

together with the array of subcomplexes given by the total singular complex of $E^{Q^+}|\tau$ for all closed subsets Q^+ of $P^+(\sigma)$ and all faces τ of σ and the *dual classes* for *Morse layers* defined below.

Suppose that $Q^+ = A\coprod\{x\}$ where A is a closed subset of Q^+ and x is a maximal element. Then the *layer* of the filtered chain complex corresponding to (A, x, τ) is the subquotient complex

$$\frac{S_*(E^{Q^+}|\tau;\rho)}{S_*(E^A|\tau;\rho)+\sum S_*(E^{Q^+}|\partial_i\tau;\rho)}. \tag{2.13}$$

Depending on whether x is a Morse component of $\Sigma(f|\sigma)$ or a component containing b-d points, we call (2.13) a *Morse* or *birth-death layer.*

The birth-death layers are all acyclic. Morse layers are, however, nontrivial in exactly one degree equal to the index i of x plus the dimension d of τ. In degree $i+d$ the Morse layer (2.13) has homology isomorphic to R. The generator is dual to the $d+i$ dimensional cohomology class given by the ascending disks of the Morse critical point of f_t corresponding to x where t ranges through all interior points of τ. Note that the orientation of the function f gives a $\mathbb{Z}$–orientation of the normal bundle of these ascending disks. If we choose paths from the base point of E to each component of $\Sigma(f|\sigma)$ we get an R–orientation, i.e., a generator of the $d+i$ cohomology of the layer (2.13). We call this cohomology generator the *dual class.* It depends of the choice of paths from x to the base point of E.

The following theorem is the theorem from [**IK95**] on the existence of a "minimal subcomplex" of a filtered chain complex reinterpreted in [**Igu02**] and expressed in the language of [**Bro59**].

THEOREM 2.14 (existence of twisting cochains). *There exists a Δ^k–family of chain complexes $(P(\sigma), \phi(\sigma))$ which is uniquely determined up to simplicial homotopy by the fact that its total complex (the twisted tensor product),*

$$C_*(\Delta^k)\otimes_{\phi(\sigma)} P(\sigma)R$$

(Definition 2.12), together with the filtration given by closed subsets of $P(\sigma)$ and faces of σ, is filtered quasi-isomorphic to the total singular complex of the pair $(E|\sigma, \partial_0 E|\sigma)$ with twisted R–coefficients and so that the basis element of each layer in the twisted tensor product is dual to the dual class of the filtered chain complex.

Here, *filtered quasi-isomorphic* means there is a filtration preserving chain map which induces a quasi-isomorphism on every subcomplex (and thus on every layer). The twisted tensor product is *minimal* in the sense that every Morse layer is free of rank 1, every birth-death layer is zero as is the base layer corresponding to the empty set.

REMARK 2.15 (definition of FR–torsion is an algebraic problem). Note that Brown's theorem (that the twisted tensor product gives the homology of the total space) holds by construction over each simplex and thus, by Mayer-Vietoris, for the entire bundle. The higher Franz-Reidemeister torsion is the obstruction to deforming the twisted tensor product given by a fiberwise framed function on E to one given by a canonical algebraically defined twisting cochain in situations when the latter exists. Therefore, it is a purely algebraic problem to determine under what conditions higher FR–torsion is defined.

2.5. Subfunctors

A subfunctor of a monomial functor (P, ϕ) is given by a system of closed subsets of the poset functor P.

If ϕ is a twisting cochain over R subordinate to the constant poset P then any closed subset Q of P defines a twisting cochain ϕ_Q with coefficients in $\mathrm{End}(QR)$: For every vertex v, Q generates a subcomplex $(QR, \phi_Q(v))$ of $(PR, \phi(v))$. For every edge (v_0, v_1), we get a commuting diagram of chain complexes:

$$\begin{array}{ccc} (PR, \phi(v_1)) & \xrightarrow[\approx]{1+\phi(v_0,v_1)} & (PR, \phi(v_0)) \\ \subseteq\uparrow & & \subseteq\uparrow \\ (QR, \phi_Q(v_1)) & \xrightarrow[\approx]{1+\phi_Q(v_0,v_1)} & (QR, \phi_Q(v_2)) \end{array}$$

and so on. We also get a *quotient twisting cochain* ϕ/ϕ_Q subordinate to P/Q giving the quotient complexes $(PR, \phi(v))/(QR, \phi_Q(v))$ at vertices, etc.

Given a poset functor $P : \mathcal{C} \to \mathcal{P}(G)$ a *subfunctor* Q is defined to be a closed subset $Q(A)$ of $P(A)$ for each object A of $\mathcal{C}$ which is *invariant* in the sense that

$$f_*(Q(A)) = Q(B)$$

for all morphisms $f : A \to B$ in $\mathcal{C}$. We also have the *quotient functor* P/Q given by $(P/Q)(A) = P(A)/Q(A)$.

Finally, if $\xi = (P, \phi)$ is an (R, G)–monomial functor then a poset subfunctor Q of P gives a *monomial subfunctor*

$$\eta = (Q, \phi_Q)$$

of ξ and we also have the *monomial quotient functor*

$$\xi/\eta = (P/Q, \phi/\phi_Q).$$

We say that ξ is an *extension* of η by ξ/η.

We need one more definition. If a subset S of a poset P is both open and closed we call it *independent*. In that case P is a disjoint union of two closed subsets Q, S, i.e., the elements of Q are unrelated to the elements of S. We write

$$P = Q \vee S$$

(although posets have no base points). If (P, ϕ) is a monomial functor and $P = Q \vee S$ (a disjoint union of poset subfunctors) then

$$\phi = \phi_Q \oplus \phi_S$$

since upper triangular endomorphisms of PR have no cross terms (QR does not map to SR and SR does not map to QR). Therefore,

$$(P, \phi) = (Q, \phi_Q) \oplus (S, \phi_S).$$

2.6. The Whitehead category

We now need to construct two simplicial categories $\mathcal{M}_\bullet(R, G)$ and $\mathcal{W}h_\bullet(R, G)$ which will serve as the target category for the (R, G)–monomial functors defined above and for the more general (R, G)–expansion functors, respectively. The arrows are reversed from the definitions of the same terms in [**Igu02**]. We will see that this makes no problems.

DEFINITION 2.16 (Whitehead category). The *Whitehead category* $\mathcal{W}h_\bullet(R, G)$ is defined to be the simplicial category with simplicial set of objects equal to $\Delta^\bullet(R)$. A morphism $(P, \phi) \to (Q, \psi)$ in $\mathcal{W}h_k(R, G)$ is defined to be a decomposition $P = A \vee B$ and a morphism $\alpha : A \to Q$ in $\mathcal{P}(G)$ so that

(1) The restriction ϕ_A of ϕ to A corresponds to ψ under the ring isomorphism

$$\mathrm{End}_R(AR) \to \mathrm{End}_R(QR)$$

induced by α, i.e., $(\phi_A)_\# = \psi$.

(2) $B = S_1 \vee \cdots \vee S_n$ where each $S_i = \{x_i^+, x_i^-\}$ so that $\phi_0(j)(x_i^+) = x_i^-$ (i.e., $\phi_0(j)(x_i^+, 1) = (x_i^-, g)$ for some $g \in G$) for every $j \in [k]$.

Let $\mathcal{M}_\bullet(R, G)$ denote the simplicial subcategory of $\mathcal{W}h_\bullet(R, G)$ having all of the objects but only *monomial morphisms* (i.e., those with $P = A$ and $B = \emptyset$). The pairs S_i are called *collapsing pairs* in P. They are independent by assumption.

Morphisms in the Whitehead category factor uniquely:

$$(A \vee B, \phi_A \oplus \phi_B) \xrightarrow{c} (A, \phi_A) \xrightarrow{\alpha} (Q, \psi).$$

The second morphism is monomial. The first is called a *collapse* (or *inverse expansion*). (In the definition above, the morphism is an collapse if $A = Q$ and α is the identity morphism.) The induced mapping

$$\mathrm{End}_R(AR \oplus BR) \twoheadrightarrow \mathrm{End}_R(AR) \xrightarrow{\approx} \mathrm{End}_R(QR)$$

is not a ring homomorphism. The first map is given by

$$\begin{pmatrix} a & b \\ c & d \end{pmatrix} \mapsto a.$$

The second map is the isomorphism induced by α. However, we do get a homomorphisms of rings without unit on the strictly upper triangular endomorphisms because, on that set, b and c in the matrix above are zero.

REMARK 2.17 (reinterpreting monomial functors). An (R, G)–monomial functor $\xi = (P, \phi)$ on a simplicial set X can now be reinterpreted as an actual functor

$$\xi : \mathrm{simp}\, X \to \mathit{diag}\, \mathrm{simp}\, \mathcal{M}_\bullet(R, G)$$

from the category of simplices of X to the diagonal of the bicategory of simplices in the simplicial category $\mathcal{M}_\bullet(R, G)$. It takes an object $(x, [k])$ of $\mathrm{simp}\, X$ to the Δ^k family

$$\xi(x, [k]) = (P(x), \phi_\#)$$

given in (2.10) above. For each map $a : [j] \to [k]$ giving a morphism $a_* : (a^*(x), [j]) \to (x, [k])$ in $\mathrm{simp}\, X$ we get a monomial morphism

$$(P(a^*(x)), \phi_\#) \to a^\sharp(P(x), \phi_\#)$$

where $a^\sharp$ is the simplicial operator on $\Delta^\bullet(R)$ induced by a. Writing $y = a^*(x)$, we can simplify the expression to

$$g_{yx} : \xi(y) \to \xi(x)|y.$$

The functoriality of ξ can be expressed by the equations:

$$g_{zx} = g_{yx}|z \circ g_{zy}, \quad g_{xx} = id_{\xi(x)}. \tag{2.14}$$

To simplify the notation we write monomial functors simply as

$$\xi : \mathrm{simp}\, X \to \mathcal{M}_\bullet(R, G).$$

DEFINITION 2.18 (expansion functor). An (R, G)–*expansion functor* on a simplicial set X is defined to be a functor, written as

$$\xi : \operatorname{simp} X \to \mathcal{W}h_\bullet(R, G),$$

from $\operatorname{simp} X$ to the diagonal of $\operatorname{simp} \mathcal{W}h_\bullet(R, G)$ which has the form $\xi(x, [k]) = (P(x), \phi(x)) \in \Delta^k(R)$ so that each map $a : [j] \to [k]$ induces a morphism

$$g_{yx} : \xi(y, [j]) = (P(y), \phi(y)) \to a^\sharp(P(x), \phi(x)) = \xi(x, [k])|y$$

in $\mathcal{W}h_j(R, G)$, where $y = a^*(x)$, satisfying (2.14) above.

Suppose that we have a fiberwise oriented GMF $f : E \to \mathbb{R}$ and a group homomorphism

$$\rho : \pi_1 E \to G \subseteq R^\times.$$

Let $\operatorname{simp} B$ be the category of small smooth transverse simplices in B. Then, disregarding for a moment the problem of independence of b-d points, we get the *cellular chain complex functor*

$$C_*(f; \rho) : \operatorname{simp} B \to \mathcal{W}h_\bullet(R, G). \tag{2.15}$$

This is the (R, G)–*expansion functor* which sends each small k–simplex σ^k in B to the cellular chain complex of $f|\sigma$ and each face or degeneracy $(\tau, [j]) \to (\sigma, [k])$ to a morphism

$$C_*(f; \rho)(\tau) \to C_*(f; \rho)(\sigma)|\tau$$

in $\mathcal{W}h_j(R, G)$. It induces a continuous mapping

$$|C_*(f; \rho)| : B \simeq |\operatorname{simp} B| \to |\mathcal{W}h_\bullet(R, G)|.$$

Conceptually, $C_*(f; \rho)$ represents a family of based free R–complexes, one for each point in B, which locally varies by G–monomial change of basis, elementary expansions, elementary basis change and other higher chain homotopies. So expansion functors

$$\xi : \operatorname{simp} B \to \mathcal{W}h_\bullet(R, G)$$

will also be called *families of chain complexes* on B.

Let $\mathcal{W}h^h_\bullet(R, G)$ be the simplicial full subcategory of $\mathcal{W}h_\bullet(R, G)$ in which all chain complexes are acyclic. Then we have the following theorem. (Since any category is homotopy equivalent to its opposite category, the theorem remains true even though the arrows are reversed.)

THEOREM 2.19 (Igusa-Klein [**IK95**], [**Igu02**]). *There is a natural homotopy fiber sequence*

$$|\mathcal{W}h^h_\bullet(R, G)| \to \Omega^\infty \Sigma^\infty(BG_+) \to BGL(R)^+ \times \overline{\mathbb{Z}}$$

where $()^+$ is the Quillen plus construction, $()_+$ means add a disjoint base point and $\overline{\mathbb{Z}}$ is the image of $\mathbb{Z}$ in $K_0 R$.

REMARK 2.20 ($Q(BG_+) \simeq \Omega|\mathcal{S}_\bullet \mathcal{W}h_\bullet(R, G)^{op}|$). The middle term $\Omega^\infty\Sigma^\infty(BG_+)$ is the *Waldhausen K–theory* or *group completion* of (a Waldhausen category homotopy equivalent to) $\mathcal{W}h_\bullet(R, G)$ and the map from $|\mathcal{W}h^h_\bullet(R, G)|$ into it is induced by inclusion. The first term $|\mathcal{W}h^h_\bullet(R, G)|$ is already an infinite loop space so is equal to its group completion.

REMARK 2.21 (standard notation for Whitehead space). The classifying space $|\mathcal{W}h^h_\bullet(R,G)|$ corresponds to the *loop space* of the usual Whitehead space. Thus, for a connected CW complex X the linearization map $\lambda : Q(X_+) \to \mathbb{Z}[\pi_1 X]$ gives a mapping

$$\lambda_* : \Omega Wh^{diff}(X) \to |\mathcal{W}h^h_\bullet(\mathbb{Z}[\pi_1 X], \pi_1 X)|$$

from the loop space of Waldhausen's smooth Whitehead space $Wh^{diff}(X)$ to the classifying space of our Whitehead category. If we compare Waldhausen's theorem ([**Wal87**])

$$A(X) \simeq Q(X_+) \times Wh^{diff}(X)$$

with Theorem 2.19 above, we conclude that the map λ_* is a rational homotopy equivalence when $X \simeq BG$. We will not use this fact.

In the special case when $R = \mathbb{Z}, G = 1$ we conclude from Theorem 2.19 and Borel's computation of the rational K–theory of $\mathbb{Z}$ ([**Bor74**]) that

COROLLARY 2.22 (rational homotopy of $\mathcal{W}h^h_\bullet(\mathbb{Z},1)$).

$$\pi_i \mathcal{W}h^h_\bullet(\mathbb{Z},1) \otimes \mathbb{Q} = \begin{cases} \mathbb{Q} & \text{if } i = 4k \text{ with } k > 0, \\ 0 & \text{otherwise.} \end{cases}$$

This implies that the real cohomology ring of $\mathcal{W}h^h_\bullet(\mathbb{Z},1)$ is a polynomial ring in generators

$$\tau_{2k} \in H^{4k}(\mathcal{W}h^h_\bullet(\mathbb{Z},1);\mathbb{R}).$$

A formula for these classes is given (more generally) as follows.

2.7. Definition in acyclic case

First we need to reduce to the two–index case using the *two–index theorem.*

THEOREM 2.23 (two–index theorem). *For any ring R, subgroup G of $R^\times$ and integers $0 \le i < j$, the simplicial full subcategory $\mathcal{W}h^{h[i,j]}_\bullet(R,G)$ of $\mathcal{W}h^h_\bullet(R,G)$ consisting of acyclic R–complexes with basis posets having elements only in degrees $i, i+1, \cdots, j$ is a deformation retract.*

Consequently, we have:

$$\mathcal{W}h^{h[0,1]}_\bullet(R,G) \simeq \mathcal{W}h^h_\bullet(R,G).$$

We may regard $\mathcal{W}h^{h[0,1]}_\bullet(R,G)$ as a space of invertible matrices with coefficients in R which are only well-defined up to left and right multiplication by G–monomial matrices.

Now restrict to the case $(R,G) = (M_r(\mathbb{C}), U(r))$. Then

$$\mathcal{W}h^{h[0,1]}_\bullet(R,G) = \mathcal{W}h^{h[0,1]}_\bullet(M_r(\mathbb{C}), U(r)) \tag{2.16}$$

is a space of invertible complex matrices which are well-defined up to left and right multiplication by unitary matrices. We modify the definition so that every k–simplex is actually equal to a smooth family of invertible matrices over $\mathbb{C}$ and morphisms multiply these by unitary matrices. We will come back to this point in the next section.

Next, we use the *Kamber-Tondeur form* ([**KT74**], [**Dup76**]). This gives a simplicial $2k$–cochain c_{2k} on the smooth version of the simplicial category (2.16) given on a smooth $2k$–simplex $f_t, t \in \Delta^k$, by

$$c_{2k}(f_t) = \frac{1}{2i^k(2k+1)!} \int_{(t,u)\in\Delta^{2k}\times I} \mathrm{Tr}((h_t^{-u} dh_t^u)^{2k+1}) \tag{2.17}$$

where $h_t = f_t f_t^*$. By adding certain polynomial correction terms to c_{2k} we get a cocycle D_{2k}. We do not need to know what these correction terms are because the higher FR–torsion is a rational multiple of a polylogarithm and polylogarithms are linearly independent from polynomials. ([**Igu02**], Lemma 7.7.2.)

The result is that we have well-defined real cohomology classes

$$\tau_k = [D_{2k}] \in H^{2k}(\mathcal{W}h_\bullet^h(M_r(\mathbb{C}), U(r)); \mathbb{R}) \tag{2.18}$$

which we call the *universal Franz-Reidemeister torsion invariants.* (It is independent of r in a natural sense.) To get invariants of smooth bundles we proceed as follows.

Let $(E, \partial_0 E) \to B$ be a smooth bundle pair with fiber $(M, \partial_0 M)$ and let $\mathcal{F}$ be a Hermitian coefficient system on E. This is given by a unitary representation $\rho : \pi_1 E \to U(r)$. Suppose further that

$$H_*(M, \partial_0 M; \mathcal{F}) = 0. \tag{2.19}$$

Suppose there is a fiberwise framed function

$$f : (E, \partial_0 E) \to (I, 0)$$

which is unique up to framed homotopy. The Framed Function Theorem (2.4) says this holds if $\dim M > \dim B$. Then, setting aside a problem with the b-d points, the cellular chain complex functor (2.15) of f with coefficients in $\mathcal{F}$ is an expansion functor

$$C_*(f; \mathcal{F}) : \mathrm{simp}\, B \to \mathcal{W}h_\bullet^h(M_r(\mathbb{C}), U(r)).$$

So, we can pull back the universal higher FR–torsion invariant τ_k to obtain a well-defined cohomology class

$$\tau_k(E, \partial_0 E; \mathcal{F}) = C_*(f; \mathcal{F})^*(\tau_k) \in H^{2k}(B; \mathbb{R}).$$

This is called the *higher Franz-Reidemeister torsion invariant* of the pair $(E, \partial_0 E)$ with coefficients in $\mathcal{F}$.

2.8. Families of matrices as flat superconnections

We will explain what we mean by a smooth family of invertible matrices. By differentiating the components of the structure we will obtain what is known as a flat $\mathbb{Z}$–graded superconnection of total degree 1 [**BL95**]. (See also Quillen's original article on superconnections [**Qui85**].) This is explained in detail in [**Igu02**] although the superconnection formalism is not mentioned there. The relation to superconnections was observed by Goette [**Goe**].

DEFINITION 2.24 (smooth category of invertible matrices). The *smooth category of invertible matrices* is defined to be the simplicial category $\mathcal{W}_\bullet^{diff}(\mathbb{C}^r, U(r))$ whose objects in degree k are triples (P, C_*, E) where

(1) $P = P_0 \coprod P_1$ is a graded poset with $|P_0| = |P_1| = n$ for some $n \geq 0$.

(2) $C_*(t) = (P\mathbb{C} \otimes \mathbb{C}^r, \phi_0(t)), t \in \Delta^k$ is a smooth Δ^k–family of acyclic chain complexes of the form

$$P_1\mathbb{C} \otimes \mathbb{C}^r \xrightarrow{\phi_0(t)} P_0\mathbb{C} \otimes \mathbb{C}^r$$

where the boundary map $\phi_0(t)$ is strictly upper triangular and gives a smooth mapping

$$\phi_0 : \Delta^k \to GL_{nr}(\mathbb{C}).$$

(3) $E(s,t)$ is a smooth family of upper triangular chain isormorphisms

$$E(s,t) : C_*(t) \xrightarrow{\approx} C_*(s)$$

for $(s,t) \in \Delta^k \times \Delta^k$ so that $E(t,t) = I$ (the identity map) and $E(s,t) - I$ is strictly upper triangular for all s, t.

(4) There exists a unique family of strictly upper triangular chain homotopies

$$S(s,t,u) : E(s,u) \simeq E(s,t)E(t,u)$$

for $(s,t,u) \in \Delta^k \times \Delta^k \times \Delta^k$ so that $S(s,t,u) = 0$ if s,t,u are not distinct.

The morphisms will be given later.

REMARK 2.25 (correspondence with Morse theory). (1) The poset P is the set components in a Δ^k family of Morse functions with critical points in two indices. P_0, P_1 are the subsets of P corresponding to critical points of fixed indices $m, m+1$. The partial ordering of P must respect the "trajectory order," i.e., $x < y$ if there is a trajectory of $-\nabla f_t$ going from y down to x for some $t \in \Delta^k$.

(2) The incidence matrix of f_t counts trajectories of $-\nabla f_t$ going from elements of P_1 to elements of P_0. This is a discontinuous function of $t \in \Delta^k$. We choose a smoothing which still respects the partial ordering.

(3) In a one–parameter family of functions there will be incidences between elements of P_i with P_i for $i = 0, 1$. The off-diagonal entries of the matrix $E_i(s,r)$ count the number of such occurrence for f_t as t travels in a straight line from s to r in Δ^k. $E_0(s,t), E_1(s,t)$ together form a chain isomorphism. Again we apply a smoothing operation.

(4) In a two–parameter family of functions there are trajectories of $-\nabla f_t$ from some $x_i \in P_0$ down to some $y_j \in P_1$. This can only occur if the ij–entry of the incidence matrix ϕ_0 is zero. The ji–entry of the matrix $S(s,t,u)$ counts the number of times this occurs in the triangle with corners s, t, u in Δ^k (with a smoothing operation applied). For the family of functions f_t the incidence of x_i over y_j "exchanges" row operations and column operations. This is because, when the m–cell x_i slides over the $m+1$–cell y_j, it also will also slide over the m–cells incident under y_j and any $m+1$–cell incident over x will now slide over y_j. See [**HW73**] for more details about the exchange relation

The chain homotopy $S(s,t,u)$ is part of the structure of a smooth family of matrices. However, when the matrices are invertible, $S(s,t,u)$ is uniquely determined. So, we list its existence as a condition instead of as an additional structure.

REMARK 2.26 (homotopy equivalence with the Whitehead category). The simplicial category $\mathcal{W}_\bullet^{diff}(\mathbb{C}^r, U(r))$ and its continuous counterpart $\mathcal{W}_\bullet^{cont}(\mathbb{C}^r, U(r))$

are both homotopy equivalent to the combinatorial space $\mathcal{W}h_\bullet^{h[0,1]}(M_r(\mathbb{C}), U(r))$. This is Theorem 2.5.11(a) in [**Igu02**]. The homotopy equivalence is given by forgetful functors:

$$\mathcal{W}_\bullet^{diff}(\mathbb{C}^r, U(r)) \xrightarrow{\simeq} \mathcal{W}_\bullet^{cont}(\mathbb{C}^r, U(r)) \xrightarrow{\simeq} \mathcal{W}h_\bullet^{h[0,1]}(M_r(\mathbb{C}), U(r)).$$

By the two index theorem (2.23) these are all homotopy equivalent to the Whitehead category $\mathcal{W}h_\bullet^h(M_r(\mathbb{C}), U(r))$.

A smooth family of chain complexes (P, C_*) is equivalent to a flat $\mathbb{Z}$-graded superconnection on $P\mathbb{C}\otimes\mathbb{C}^r$ of total degree 1. This was observed by Goette [**Goe**]. The idea is to differentiate the structure maps. In our case there are only three structure maps $\phi_0(t), E(s,t)$ and $S(s,t,u)$.

Since $E(s,t) : C_*(t) \to C_*(s)$ is a chain isomorphism we have:

$$E(s,t)\phi_0(t) = \phi_0(s)E(s,t).$$

Differentiating both sides with respect to s and put $s = t$ we get

$$(d_1E)\phi_0 = d\phi_0 + \phi_0 d_1 E$$

where d_1 means differential with respect to the first variable (so $d_1E + d_2E = 0$ since $E(t,t) = I$ is constant). If we denote the matrix 1–form d_1E by ϕ_1 we get

$$d\phi_0 = \phi_1\phi_0 - \phi_0\phi_1. \tag{2.20}$$

The chain maps $E(s,t)$ are only required to form a functor up to homotopy by $S(s,t,u)$:

$$E(s,t)E(t,u) = E(s,u) + \phi_0(s)S(s,t,u) + S(s,t,u)\phi_0(u).$$

When this equation is differentiated with respect to s and t and the variables are equated, we get:

$$d_1d_2E + d_1E\, d_1E = \phi_0 d_1d_2S + (d_1d_2S)\phi_0.$$

Denoting the matrix 2–form d_1d_2S by ϕ_2 this becomes:

$$-d\phi_1 = \phi_0\phi_2 - \phi_1^2 + \phi_2\phi_0. \tag{2.21}$$

LEMMA 2.27 (derivative of top form). *The following equation follows from (2.20) and (2.21).*

$$d\phi_2 = \phi_1\phi_2 - \phi_2\phi_1. \tag{2.22}$$

PROOF. Differentiate (2.21) and replace $d\phi_0$ and $d\phi_1$ by their values as given in (2.20) and (2.21). The result can then be written as

$$Z\phi_0 = \phi_0 Z$$

where

$$Z = d\phi_2 - \phi_1\phi_2 + \phi_2\phi_1.$$

But Z maps C_0 to C_1 so we must have $Z\phi_0 = \phi_0 Z = 0$ which implies that $Z = 0$. □

A $\mathbb{Z}$-graded superconnection $D = d + A_0 + A_1 + A_2$ of total degree 1 is *flat* if

$$-dA = A^2.$$

However, Quillen's notation [**Qui85**] places the forms on the left of the matrices. If we use the supercommutator rules to put A^2 in *normal form* by moving both forms to the left of both matrices (See [**Igu**] for more details about this point) we get the following.

(1) $dA_0 = A_0A_1 - A_1A_0$
(2) $-dA_1 = A_0A_2 + A_1^2 + A_2A_0$
(3) $dA_2 = -A_1A_2 + A_2A_1$

This matches (2.20), (2.21) and (2.22) if we let $A_0 = \phi_0, A_1 = -\phi_1$ and $A_2 = -\phi_2$. So, we get the following.

PROPOSITION 2.28 (families of matrices give superconnections).

$$D = d + \phi_0 - \phi_1 - \phi_2$$

is a flat superconnection on Δ^k.

REMARK 2.29 (notation change). In [**Igu02**] the notation is different. We used the symbols X, U, V, W, f to denote:

(1) $f = \phi_0$. (This is an isomorphism $C_1 \xrightarrow{\approx} C_0$.)
(2) $(U, V) = \phi_1$. (So $U = \phi_1|C_0$, $V = \phi_1|C_1$.)
(3) $W = \phi_2\phi_0$. (Or $\phi_2 = Wf^{-1}$.)
(4) $X = \phi_0^{-1}d\phi_0$. (Or $d\phi_0 = fX$.)

REMARK 2.30 (role of superconnections in definition of higher FR–torsion). The $2k$–cocycle D_{2k} which gives the higher torsion is equal to C_{2k} given in (2.17) plus a linear combination of integrals of traces of products of X, V, W which are endomorphisms of C_1. (The term U is an endomorphism of C_0 and occurs only in the form $f^{-1}Uf = -X - V$.)

Finally, we give the definition of the morphisms in the category $\mathcal{W}_\bullet^{diff}(\mathbb{C}^r, U(r))$. They are just as in the Whitehead category. A morphism $(P, \phi) \to (Q, \psi)$ is a decomposition $P = A \vee B$ together with a morphism $\alpha : A \to Q$ in $\mathcal{P}(U(r))$ so that

(1) For all $t \in \Delta^k$, the restriction $\phi_A(t)$ of $\phi(t)$ to A corresponds to ψ under the isomorphism
$$\Omega(\Delta^k, \mathrm{End}(A\mathbb{C} \otimes \mathbb{C}^r)) \cong \Omega(\Delta^k, \mathrm{End}(Q\mathbb{C} \otimes \mathbb{C}^r))$$
induced by α.
(2) $B = S_1 \vee \cdots \vee S_n$ where each S_i is a collapsing pair x^+, x^- with constant $\phi_0(t) \in U(r)$ and trivial ϕ_1 and ϕ_2.

2.9. Independence of birth-death points

We will discuss briefly a problem with the b-d points and how it is solved. The problem is that the definition of an expansion functor

$$\mathrm{simp}\, B \to \mathcal{W}h_\bullet(R, G)$$

assumes that collapsing pairs are independent. The independence of b-d points would automatically imply this algebraic condition for the cellular chain functor constructed above. However, even if b-d points are not independent, the condition can still be achieved in a canonical algebraic way.

A b-d point x of M_t is called *independent* if x is not incident to any other critical point of f_t. In other words, $\{x\}$ is an independent set in the poset $P^+(t) = \Sigma(f_t)$. Suppose for a moment that all b-d points are independent. Since independence is an open condition, this would imply that, for any small smooth transverse simplex σ in B, any component $x \in P^+(\sigma) - P(\sigma)$ (i.e., x is a component of $\Sigma(f|\sigma)$ containing a b-d point), will be independent in the poset $P^+(\sigma)$.

Since $x \notin P(\sigma)$, it does not directly appear as part of the expansion functor $\xi(\sigma) = (P(\sigma), \phi_\sharp(\sigma))$. However, it may produce a collapsing pair on a face τ of σ. This happens if $x|\tau$ has no b-d points. Then $x|\tau$ is necessarily a disjoint union of two Morse components x^+, x^- which form a collapsing pair. The independence of x implies that $S = \{x^+, x^-\}$ is independent in $P(\tau)$ as required by Definition 2.16.

The condition of independence of b-d points does not hold in general. For example, if M_t is a closed manifold, it is impossible for b-d points to be independent. One way to get around this problem is by a combination of positive and negative suspension. However, this produces manifold bundles with corners which complicates the geometry. The method that we prefer is algebraic.

In [**Igu02**], a rather complicated explicit algebraic formula is given to deform a nonindependent collapsing pair into an independent one. The reason it is complicated is because this deformation must be carried out simultaneously and in a continuous way on all collapsing pairs which appear in the expansion functor ξ.

Here, we will use the local splitting lemma (Lemma 3.13 below) which tells us that such a deformation exists and is unique up to homotopy since collapsing pairs form an acyclic subquotient complex.

The preferred solution to our problem is as follows. We do not assume independence of b-d points. Instead we choose the twisting cochains on the simplices τ which are faces of some σ as described above so that the collapsing pairs are algebraically independent. Then we alter the partial ordering on $P(\tau)$ so that each collapsing pair is poset independent.

We also need to alter the partial ordering of $P^+(\sigma)$ in such a way that collapsing pairs and "expanded pairs" act in unison. Here, a *collapsing pair* in $P^+(\tau)$ is a pair of Morse components x^+, x^- which collapse (to one b-d component) in another simplex σ of which τ is a face. A pair of Morse components x^+, x^- in $P(\sigma)$ which restricts to a collapsing pair on some face τ is an *expanded pair*. To make these pairs *act in unison* we increase the partial ordering so $x^- < y$ implies $x^+ \leq y$ and $z < x^+$ implies $z \leq x^-$. The poset $P^+(\sigma)$ is used to construct the filtered chain complex of f over σ. We are just discarding those subcomplexes which separate the members of a collapsing or expanded pair, i.e., we combined these into one layer.

With this solution of the independence of b-d point problem, the only topological problem we have in the definition of higher FR–torsion is the existence of fiberwise framed functions. For the moment, we assume that the higher FR–torsion is only defined when the fiber dimension is large. Later, using properties of the higher torsion, we will be able to define the higher torsion of $(E, \partial_0 E)$ to be half the torsion of $(E \times S^{2N}, \partial_0 \times S^{2N})$ for N large. (See Remark 3.29.)

In the absolute case, when $\partial_0 E$ is empty ($\partial E = \partial_1 E$), we can define the higher FR–torsion of E to be that of $E \times D^N$ (with corners rounded). This is well-defined by the positive suspension lemma proved below.

2.10. Positive suspension lemma

Suppose that

$$f : (E, \partial E) \to (I, 1)$$

is a fiberwise oriented GMF. Let $C_*(f; \rho)$ be the (R, G)–expansion functor for f with coefficients in a representation $\rho : \pi_1 E \to G \subseteq R^\times$.

DEFINITION 2.31 (positive suspension). Let $\pi : D \to E$ be the unit disk bundle of an m–plane bundle on E. Then the *positive suspension*

$$h = \sigma_+ f : D \to \mathbb{R}$$

of f is the map given by

$$h(y) = f(x) + \|y\|^2$$

where $x = \pi(y)$.

Then $\sigma_+ f$ is an oriented fiberwise GMF with the same cellular chain complex as f. We will prove this obvious statement. The proof will explain something about the definition of $C(f;\rho)$ and will also give the definition of higher torsion for low dimensional fibers as promised at the end of the last section.

LEMMA 2.32 (positive suspension lemma). *The (R,G)–expansion functors of f and $\sigma_+ f$ are equal.*

$$C_*(\sigma_+ f;\rho) = C_*(f) : \operatorname{simp} B \to \mathcal{W}h_\bullet(R,G).$$

REMARK 2.33 (vertical metric on D). To define $C_*(\sigma_+ f;\rho)$ we need a vertical metric on D as a bundle over B. As an $O(m)$–disk bundle, D has a vertical metric over E. Choose a smooth horizontal distribution for D over E given by an $O(m)$–connection. Then the horizontal directional derivatives of $\|y\|^2$ will be zero. At each point $y \in D$ let K_y be the kernel of the composition

$$T_y D \xrightarrow{D\pi} T_x E \xrightarrow{Dp} T_{p(x)} B$$

restricted to the horizontal plane at y. Then K_y maps isomorphically to the vertical tangent plane of E at x:

$$D\pi : K_y \xrightarrow{\approx} T_x^v E. \tag{2.23}$$

Pull back the vertical metric on E (the one which was used to define $C(f;\rho)$) and make K_y perpendicular to the fiber of D over x. This defines a vertical metric for D over B.

PROOF. By definition of the vertical metric of D over B, the gradient of $h = \sigma_+ f$ decomposes into vertical and horizontal components

$$\nabla h(y) = 2y + \pi^* \nabla f(x) \tag{2.24}$$

where π^* is the inverse of $D\pi$ in (2.23). Therefore, the critical points of h lie in the zero section of E in D and coincide with the critical points of f (if we identify E with its zero section in D). Furthermore, $\nabla h = \nabla f$ on this zero section. Thus the inclusion of the zero section is filtration preserving and thus induces a morphism of filtered chain complexes from the total singular complex of $E|\sigma$ to that of $D|\sigma$ for every small transverse simplex σ in B. By looking at the layers, we see that this inclusion map is a filtered quasi-isomorphism preserving the dual classes. So $C(f;\rho)$ and $C_*(h;\rho)$ can be taken to be equal. (They are only well defined up to simplicial homotopy.) □

This lemma can be used to define the cellular chain complex functor on any smooth bundle $E \to B$ with $\partial_0 E = \emptyset$.

THEOREM 2.34 (cellular chain complex for $\partial_0 E$ empty). *If $E \to B$ is a smooth bundle and ρ is a representation of $\pi_1 E$ in $G \subseteq R^\times$ then we get an (R, G)–expansion functor*

$$\xi : \operatorname{simp} B \to \mathcal{W}h_\bullet(R, G)$$

which is well-defined up to homotopy by taking $\xi = C_*(f; \rho)$ where f is a fiberwise framed function on $E \times D^N$ for N large.

PROOF. The statement is that ξ does not depend on N and f (up to homotopy). So suppose that D is another disk bundle over E and $g : D \to \mathbb{R}$ is a fiberwise framed function whose gradient point outward on all faces of the boundary. Then we get two positive suspensions

$$\sigma_+ f, \sigma_+ g : D \times D^N \to \mathbb{R}.$$

By the Framed Function Theorem, these are homotopic, so

$$C_*(f; \rho) = C_*(\sigma_+ f; \rho) \simeq C_*(\sigma_+ g; \rho) = C_*(g; \rho).$$

□

Theorem 2.34 can be used to define higher FR–torsion when the fiber F is acyclic with respect to a unitary representation of $\pi_1 E$. This can be extended to the "upper triangular" case by an algebraic construction which we call the "canonical cone."

2.11. Definition in upper triangular case

The following lemma allows us to generalize the condition (2.19) under which τ_k is defined. We need one definition. If K is a field and π is a group, we say that a finite dimensional $K[\pi]$–module M is *upper triangular* if it contains a filtration by $K[\pi]$–submodules (i.e., a composition series)

$$M = M_n \supseteq M_{n-1} \supseteq \cdots \supseteq 0$$

so that π acts trivially on the subquotients M_i / M_{i-1}.

LEMMA 2.35 (canonical cone). *Let ξ be a family of chain complexes*

$$\xi : \operatorname{simp} B \to \mathcal{W}h_\bullet(R, G)$$

satisfying either of the following two conditions where $v \in B$ is any base point.

1. *The homology groups $H_n(\xi(v); R)$ are projective R–modules with trivial $\pi_1 B$ action.*
2. *R is a field and each $H_n(\xi(v); R)$ is an upper triangular $\pi_1 B$–module.*

Then there exists a family of acyclic chain complexes

$$C(\xi) : \operatorname{simp} B \to \mathcal{W}h^h_\bullet(R, G)$$

containing ξ as a subfunctor which is canonically defined up to homotopy.

We call $C(\xi)$ the *canonical cone* of ξ.

REMARK 2.36 (canonical cone = fiberwise mapping cone). In **[Igu02]** $C(\xi)$ is called the *fiberwise (mapping) cone.* It is constructed as the fiberwise mapping cone of a fiberwise chain homotopy equivalence $f : \widetilde{H}(\xi, Q) \to \xi$ where

$$\widetilde{H}(\xi, Q) : \operatorname{simp} B \to \mathcal{W}h^h_\bullet(R, G) \tag{2.25}$$

is the "twisted homology bundle" which is explained in the proof below. Basically, $\widetilde{H}(\xi, Q)$ is a family of based free chain complexes with basis Q and zero boundary map which is chain homotopy equivalent to ξ. I.e., it is the homology of ξ at each $t \in B$. However, it is not constant since it has the higher homotopies required to make it fiberwise chain homotopy equivalent to ξ. The mapping f is not a morphism in the category $\mathcal{W}h^h_\bullet(R, G)$. Therefore, ξ is not homotopic to $\widetilde{H}(\xi, Q)$ in general. (Equivalently, $C(\xi)$ is not null-homotopic in general.)

PROOF. The existence of the canonical (fiberwise) cone is proved in [**Igu02**], Theorem 3.4.9 in the free subcase of Case 1 and in Theorem 3.4.11 for Case 2. The uniqueness up to homotopy is proved in Section 3.5 of [**Igu02**]. The ideas, however, are very simple and we will outline them here.

First, we need to reduce Case 1 to the free case. To do this we simply take direct sum with a constant finitely generated free chain complex with complementary projective homology. This is possible since the alternating sum of homology groups is stably free as a virtual R–module. By *constant* we mean independent of $t \in B$, i.e., the poset P is constant and all higher homotopies are zero.

(*Existence*) We now have free homology with the property that a basis Q can be chosen so that at least one basis element in the top homology group $H_n(\xi)$ is fixed by $\pi_1 B$. By induction it is enough to kill this element. To do this we introduce a new basis element in degree $n+1$ which is maximal. We just need to define the boundary of this element. Over each vertex $b \in B$ choose any n–chain in $\xi(b)$ representing the invariant generator of $H_n(\xi)$. The obstruction to extending this over each 1–simplex e of B lies in $H_{n+1}(\xi(e)) = 0$ and lies in $H_{n+k}(\xi(\sigma)) = 0$ for k–simplices σ. I.e., there is no obstruction and we can construct a family of chain complexes ξ' containing ξ as a subfunctor with smaller homology. Iteration gives an acyclic chain complex $C(\xi)$.

The quotient $C(\xi)/\xi$ has basis poset Q equal to the homology basis. The boundary maps are all zero in the quotient complex. The desuspension of the quotient $\Sigma^{-1}(C(\xi)/\xi)$ is denoted $\widetilde{H}(\xi, Q)$ in [**Igu02**] and called the *twisted homology bundle* of ξ. In [**Igu02**], $C(\xi)$ is constructed as the fiberwise mapping cone of a fiberwise homotopy equivalence $\widetilde{H}(\xi, Q) \to \xi$.

(*Uniqueness*) Working one generator at a time, the obstruction theory tells us that the above construction is unique up to homotopy and thus up to simplicial homotopy of the functor $C(\xi)$ if the homology basis Q is fixed. Elementary basis change is also a simplicial homotopy of $C(\xi)$. Any other change of coordinates can be given by an elementary change of coordinates on $\xi \oplus \alpha$ together with a change of coordinates on β where α, β are constant maps $\operatorname{simp} B \to \mathcal{W}h^h_\bullet(R, G)$ with image in the zero component. Since $C(\xi \oplus \alpha \oplus \beta) \simeq C(\xi)$ the uniqueness of $C(\xi)$ up to homotopy follows. For details see [**Igu02**], Theorem 3.5.10 for Case 1 and Theorem 3.5.11 for Case 2. □

THEOREM 2.37 (when τ_k is defined). *Suppose that $(E, \partial_0 E) \to B$ is a smooth bundle with fiber $(M, \partial_0 M)$ and let $\mathcal{F}$ be an r–dimensional Hermitian coefficient system over E. Suppose also that either*

(1) *$\pi_1 B$ acts trivially on $H_*(M, \partial_0 M; \mathcal{F})$ or*
(2) *$r = 1$ and $H_n(M, \partial_0 M; \mathcal{F})$ is an upper triangular $\pi_1 B$–module for all $n \geq 0$.*

Then the higher Franz-Reidemeister torsion invariants

$$\tau_k(E, \partial_0 E; \mathcal{F}) \in H^{2k}(B; \mathbb{R})$$

are defined.

REMARK 2.38 (how τ_k is defined). These higher torsion invariants are given by pulling back the universal invariants (2.18) along the canonical cone

$$C(C_*(f)) : \operatorname{simp} B \to \mathcal{W}h^h_\bullet(M_r(\mathbb{C}), U(r))$$

of the cellular chain complex $C_*(f)$ associated to a fiberwise framed function

$$f : (E, \partial_0 E) \to (I, 0).$$

If f does not exist we can take the product with a large even dimensional sphere and divide by 2:

$$\tau_k(E, \partial_0 E; \mathcal{F}) = \frac{1}{2}\tau_k(E \times S^{2N}, \partial_0 E \times S^{2N}; \pi^*\mathcal{F}).$$

Here $\pi^*(\mathcal{F})$ is the pullback along $\pi : E \times S^{2N} \to E$ of the Hermitian coefficient system $\mathcal{F}$. (See Lemma 3.28 below.)

CHAPTER 3

Properties of higher FR–torsion

We go over the basic properties of higher Franz-Reidemeister torsion. Properties stated without proof are proved in [**Igu02**].

3.1. Basic properties

First we state the obvious.

PROPOSITION 3.1 (torsion of trivial bundle). If $E = \partial_0 E \times I$ then the relative torsion with respect to any coefficient system is defined and equal to zero:

$$\tau_k(\partial_0 E \times I, \partial_0 E; \mathcal{F}) = 0.$$

The next property follows immediately from the definition of higher FR–torsion and the positive suspension lemma (Lemma 2.32).

THEOREM 3.2 (positive suspension theorem). *If the higher FR–torsion invariant $\tau_k(E;\mathcal{F})$ is defined then for any linear disk bundle $q : D \to E$ we have*

$$\tau_k(D; q^*\mathcal{F}) = \tau_k(E; \mathcal{F}).$$

I.e., the higher FR–torsion of D as a bundle over B is defined and equal to the higher torsion of E.

Another obvious fact is the naturality of torsion.

PROPOSITION 3.3 (naturality of torsion). The higher FR–torsion invariants are natural on those smooth bundles $E \to B$ on which they are defined, i.e., if $E \to B$ is a smooth bundle and $f : B' \to B$ is a smooth mapping then the torsion of the pullback $f^*(E)$ of E is the pullback of the torsion:

$$\tau_k(f^*(E), f^*(\partial_0 E); f^*(\mathcal{F}) = f^*(\tau_k(E, \partial_0 E; \mathcal{F}))$$

assuming that $\tau_k(E, \partial_0 E; \mathcal{F})$ is defined.

The naturality of higher FR–torsion extends to the case when the coefficient sheaf $f^*(\mathcal{F})$ is replaced by a locally equivalent sheaf. When we say that two Hermitian coefficient systems $\mathcal{F}$, $\mathcal{F}'$ on E are *locally equivalent* we mean that $\mathcal{F}$ and $\mathcal{F}'$ are unitarily equivalent on $E|\sigma$ for every small simplex σ in B. For example, if $\mathcal{F}$ is the pullback of a coefficient system on B then it is locally equivalent to the trivial coefficient sheaf with the same fiber. We only need this in the case when the fiber is acyclic.

PROPOSITION 3.4 (locally equivalent coefficients). Suppose that $\mathcal{F}$ and $\mathcal{F}'$ are locally equivalent Hermitian coefficient systems on E and $H_*(M;\mathcal{F}) = H_*(M;\mathcal{F}') = 0$. Then

$$\tau_k(E;\mathcal{F}) = \tau_k(E;\mathcal{F}').$$

PROOF. The reason is simple. Higher FR–torsion is defined locally and is invariant under unitary change of bases. First, we have to deform the cellular chain complex functor into two indices. This can be done over each simplex of B. Thus, the results will still be locally equivalent. This means we have families of complex matrices which are conjugate by unitary matrices, i.e., $g_t' = Ug_tV$ where U, V are fixed unitary matrices (depending only on the simplex σ).

But the higher FR–torsion is given by integrals of traces of products of g_t, g_t^* and their inverses and derivatives $(d(g_tg_t^*)^u)$. These are always arranged in a natural way so that the unitary matrices U, V cancel and do not appear when we replace g_t with g_t'. Thus the trace terms are unchanged. So their integrals are equal. □

Another basic fact that we need in this paper is the *involution property.* This relates the higher FR–torsion invariant given by a fiberwise GMF f with that of its negative $-f$. We need to assume that E is *fiberwise oriented* in the sense that M is an oriented manifold and the action of $\pi_1 B$ on M preserves the orientation (otherwise, $-f$ would not be fiberwise oriented).

THEOREM 3.5 (Involution property). *Suppose that $E \to B$ is a fiberwise oriented smooth bundle and $\mathcal{F}$ is a Hermitian coefficient system on E. Let*

$$f : (E, \partial_0 E, \partial_1 E) \to (I, 0, 1)$$

be any fiberwise oriented GMF. Then the higher torsion invariants associated to f and $-f$, if defined, differ by the sign $(-1)^{n-1}$ where $n = \dim M$:

$$\tau_k(CC_*(f); \mathcal{F}) = (-1)^{n-1}\tau_k(CC_*(-f); \mathcal{F}).$$

PROOF. We outline the proof given in [**Igu02**], Theorem 5.8.1. The proof has two parts

(1) First consider the case $n = 1$. Geometrically, this corresponds to the case when M is a circle. Algebraically, we can imagine that we have a family of invertible matrices parametrized by B. The operation of *inverse conjugate transpose* is an involution on $GL_n(\mathbb{C})$ which reverses the sign of the Kamber-Tondeur form. Since inversion acts as -1 on additive cohomology classes, *conjugate transpose* does not change the sign of the higher torsion so the theorem holds for $n = 1$.
(2) In general, we use the two–index theorem. This puts the cellular chain complex of f into degrees $0, 1$ and that of $-f$ into degrees $n, n-1$. Then we desuspend the cellular chain complex of $-f$ $n-1$ times to reduce to the $n = 1$ case. By the suspension theorem this changes the sign by $(-1)^{n-1}$. □

3.2. Suspension Theorem

One of the fundamental properties of the higher torsion is that it anti-commutes with suspension. We will see later that this is geometrically obvious. (See Corollary 3.19.)

Given a chain complex (C_*, ∂), the "alternating" *suspension* $(\Sigma C, \partial^\Sigma)$ is given by $(\Sigma C)_n = C_{n-1}$ and $\partial_n^\Sigma = \partial_{n-1}$. We do not alternate the signs as in the usual definition. This version of the suspension map induces simplicial functors

$$\Sigma : \mathcal{W}h_\bullet(R, G) \to \mathcal{W}h_\bullet(R, G), \quad \Sigma : \mathcal{W}h_\bullet^h(R, G) \to \mathcal{W}h_\bullet^h(R, G)$$

(If $-1 \in G$, the usual suspension operation will also induce such a functor and the two suspension functors will be related by a natural transformation and thus be homotopic.)

THEOREM 3.6 (Suspension Theorem). *The functor*

$$\Sigma + id : \mathcal{W}h^h_\bullet(R,G) \to \mathcal{W}h^h_\bullet(R,G)$$

sending ξ to $\Sigma\xi \oplus \xi$ is null homotopic.

REMARK 3.7 (relation to Waldhausen's Additivity Theorem). This would follow from Waldhausen's Additivity Theorem [**Wal85**] except for the problem of birth-death points which occur in this construction as collapsing pairs in morphism of $\mathcal{W}h^h(R,G)$. There is a natural cofibration sequence

$$X \to C(X) \to \Sigma(X).$$

When we restrict to the subcategory $\mathcal{M}^h(R,G) = \mathcal{M}(R,G) \cap \mathcal{W}h^h(R,G)$, consisting of morphisms with no collapsing pairs, the cone $C(X)$ is naturally contractible. So $inc \vee \Sigma \simeq C \simeq *$ as maps

$$\text{“}w\mathcal{S}_\bullet\mathcal{M}^h_\bullet(R,G)\text{”} \to \text{“}w\mathcal{S}_\bullet\mathcal{W}h^h(\mathcal{R},G)\text{”} \simeq |\mathcal{W}h^h(R,G)|$$

where $\mathcal{S}_\bullet$ is Waldhausen's delooping construction and $inc : \mathcal{M}^h \to \mathcal{W}h^h$ is the inclusion functor. We put quotes “ ” since $\mathcal{M}^h, \mathcal{W}h^h$ are only homotopy equivalent to simplicial Waldhausen categories. The proof of the Suspension Theorem in [**Igu02**] rests mainly on how to handle the collapsing pairs (“expansion pairs” in [**Igu02**] since the arrows are reversed). The idea is simple. We show there exists a fixed model for the cancellation of one collapsing pair with its suspension and apply it everywhere.

Algebraic suspension is given by geometric negative suspension which we now define. This is simply the upside down version of a positive suspension.

Suppose that

$$f : (E, \partial E) \to (I, 0)$$

is a fiberwise oriented GMF. Let $C_*(f;\rho)$ be the (R,G)–expansion functor for f with coefficients in a representation $\rho : \pi_1 E \to G \subseteq R^\times$.

DEFINITION 3.8 (negative suspension). Let $\pi : D \to E$ be the unit disk bundle of an oriented m–plane bundle on E. Then the *negative suspension*

$$h = \sigma_- f : D \to \mathbb{R}$$

of f is the map given by

$$h(y) = f(x) - \|y\|^2$$

where $x = \pi(y)$.

LEMMA 3.9 (negative suspension lemma). *$\sigma_- f$ is an oriented fiberwise GMF whose cellular chain complex functor is the m–fold suspension of $C_*(f;\rho)$.*

$$C_*(\sigma_- f; \rho) = \Sigma^m C_*(f) : \operatorname{simp} B \to \mathcal{W}h_\bullet(R,G).$$

PROOF. The proof is the same as for the positive suspension lemma. The degree shift comes from the Thom isomorphism theorem.

We need to use the fact that the Thom isomorphism is given functorially at the chain level by cap product with the Thom class. The total singular complex C_* of $D|\tau$ together with the system of subcomplexes given by all faces of τ and closed subsets Q^+ of $P^+(\tau)$ is filtered chain homotopic to the subcomplex C'_* generated by simplices transverse to the zero section. On this subcomplex we have the Thom isomorphism given as a degree shifting quasi-isomorphism

$$C'_*(D|\tau, (\partial D)|\tau; \rho) \simeq \Sigma^m C_*(E|\tau, (\partial E)|\tau; \rho)$$

which preserves the filtration structure and induces a filtered equivalence. □

If the disk bundle D is trivial ($D = E \times D^m$) then the negative suspension of any framed function will be framed. Thus we can combine the above statements with the additivity property (Theorem 3.11 below) to give the following.

THEOREM 3.10 (geometric suspension theorem).

$$\tau_k(E \times D^m, \partial(E \times D^m); \pi^* \mathcal{F}) = (-1)^m \tau_k(E, \partial E; \mathcal{F})$$

when the terms are defined.

3.3. Additivity, Splitting Lemma

Higher FR–torsion satisfies two additivity properties. One comes from coefficients and the other from filtrations of chain complexes.

By additivity of coefficients we mean:

$$\tau_k(E; \mathcal{F}_1 \oplus \mathcal{F}_2) = \tau_k(E; \mathcal{F}_1) + \tau_k(E; \mathcal{F}_2).$$

This is a special case of the following theorem.

THEOREM 3.11 (additivity). *Let ξ, η be two families of acyclic based chain complexes*

$$\xi, \eta : \operatorname{simp} B \to \mathcal{W}h^h_\bullet(M_n(\mathbb{C}), U(n))$$

Then the higher torsion of their direct sum $\xi \oplus \eta$ is equal to the sum of their torsions:

$$\tau_k(\xi \oplus \eta) = \tau_k(\xi) + \tau_k(\eta).$$

These two properties follow from the additivity of the Kamber-Tondeur form under direct sum of matrices: If $f = f_1 \oplus f_2$ then $h = h_1 \oplus h_2$ and

$$\operatorname{Tr}((h^{-1}dh)^{2k+1}) = \operatorname{Tr}((h_1^{-1}dh_1)^{2k+1}) + \operatorname{Tr}((h_2^{-1}dh_2)^{2k+1}).$$

The polynomial correction terms are also additive with respect to direct sum.

One easy corollary of this is the following.

COROLLARY 3.12 (torsion of disjoint union). The higher torsion of a disjoint union is the sum of the torsions, i.e.,

$$\tau_k(E_1 \coprod E_2, \partial_0 E_1 \coprod \partial_0 E_2; \mathcal{F}_1 \coprod \mathcal{F}_2) = \tau_k(E_1, \partial_0 E_1; \mathcal{F}_1) + \tau_k(E_2, \partial_0 E_2; \mathcal{F}_2),$$

assuming the terms are defined.

The following basic lemma allows us to extend the additivity of torsion to filtered families of chain complexes. The proof is trivial but the lemma is difficult to state. Let $(P, \phi) \in \Delta^k(R)$ be a Δ^k–family of chain complexes and suppose that Q is a closed subset of P. Then the endomorphisms $\phi(\sigma)$ will not send any element of Q to any element of the complement P/Q. Thus it can be written as

$$\phi(v_*) = \begin{pmatrix} \phi_Q(\sigma) & f(\sigma) \\ 0 & \psi(\sigma) \end{pmatrix}$$

where $\psi = \phi/\phi_Q$ is the quotient twisting cochain. Note that $f(v_*)$ is a homomorphism

$$f(\sigma) : (P/Q)R \to QR.$$

LEMMA 3.13 (local splitting lemma). *Suppose that every element of Q is less than every element of P/Q. Suppose that either the subcomplex (Q, ϕ_Q) or quotient complex $(P/Q, \psi)$ is acyclic. Suppose also that $f(\sigma) = 0$ for all proper faces σ of Δ^k. Then there is a simplicial homotopy of (P, ϕ) which is the identity on Q, on P/Q and on all proper faces σ of Δ^k and so that, at the end, $f(\Delta^k)$ is also zero:*

$$(P, \phi) \simeq (Q, \phi_Q) \oplus (P/Q, \psi).$$

REMARK 3.14 (uniqueness of local splitting). It follows from the lemma, applied to the union of two homotopies, that the simplicial homotopy described in the lemma is unique up to all higher homotopies, i.e., the space of such homotopies forms a contractible space.

This lemma is meant to be used as follows. Given any (R, G)–expansion functor ξ and subfunctor η so that either η or ξ/η is acyclic, we can deform ξ into a direct sum

$$\xi \simeq \eta \oplus \xi/\eta$$

in two steps:

(1) Deform the poset functor (with expansions) so that every generator of η is less than every generator of ξ/η.
(2) Apply the lemma one simplex at a time on the domain complex to kill the incidence $f(\sigma)$ between ξ/η and η.

A jazzed-up version of this simple argument gives the following.

THEOREM 3.15 (Splitting Lemma). *Suppose that R is a ring G is a subgroup of $R^\times$ and B is a connected simplicial set with base point $v \in B_0$. Let*

$$\xi : \operatorname{simp} B \to \mathcal{W}h_\bullet(R, G)$$

be a family of based free chain complexes having a family of subcomplexes given by a functor η. Suppose that either

(1) *the homology of $\xi(v)$, $\eta(v)$ and $\xi(v)/\eta(v)$ are projective R–modules on which $\pi_1 B$ acts trivially or*
(2) *R is a field and the homology of $\xi(v)$, $\eta(v)$ and $\xi(v)/\eta(v)$ are upper triangular as $\pi_1 B$–modules.*

Then the canonical cone of ξ splits up to homotopy:

$$C(\xi) \simeq C(\eta) \oplus C(\xi/\eta).$$

In the special case when $R = M_n(\mathbb{C})$ and $G = U(n)$ this, together with additivity, implies that the higher torsion invariant is a sum:

$$\tau_k(C(\xi)) = \tau_k(C(\eta)) + \tau_k(C(\xi/\eta)).$$

3.4. Applications of the Splitting Lemma

In order to apply the Splitting Lemma we need to construct a subfunctor of the expansion functor

$$\xi : \operatorname{simp} B \to \mathcal{W}h_\bullet(R, G)$$

given by a fiberwise oriented GMF

$$f : E \to \mathbb{R}.$$

The simplest case is if there is a regular value c_t of $f_t : M_t \to \mathbb{R}$ which varies smoothly with $t \in B$. Then the critical points of f_t with critical value $< c_t$ generate a family of subcomplexes D_t of $C_*(f_t; \mathcal{F})$ giving a subfunctor

$$\eta : \operatorname{simp} B \to \mathcal{W}h_\bullet(R, G)$$

of ξ. Geometrically, E decomposes as a union of two bundles $E = E_1 \cup E_2$ along $\partial_1 E_1 = \partial_0 E_2$ where

$$\begin{aligned} E_1 &= \{x \in E \mid f_t(x) \leq c_t(t = p(x)\} \\ E_2 &= \{x \in E \mid f_t(x) \geq c_t\} \\ \partial_1 E_1 = d_0 E_2 &= \{x \in E \mid f_t(x) = c_t\} \end{aligned}$$

We refer to this situation by saying that E_2 is *stacked on top of* E_1.

COROLLARY 3.16 (stacking lemma). Suppose that E_1, E_2 are smooth bundles over B so that $\partial_1 E_1 = \partial_0 E_2$. Take the union of E_1 and E_2 along $\partial_1 E_1 = \partial_0 E_2$. Let $\mathcal{F}_1, \mathcal{F}_2$ be the restrictions to E_1, E_2 of a Hermitian coefficient system $\mathcal{F}$ on $E_1 \cup E_2$. Then

$$\tau_k(E_1 \cup E_2, \partial_0 E_1; \mathcal{F}) = \tau_k(E_1, \partial_0 E_1; \mathcal{F}_1) + \tau_k(E_2, \partial_1 E_2; \mathcal{F}_2)$$

assuming the terms are defined.

PROOF. Suppose first that there is a fiberwise framed function $f : E_1 \cup E_2 \to [0, 2]$ so that $E_i = f^{-1}[i-1, i]$. Then the Splitting Lemma says that the canonical cone of $\xi = C_*(f; \mathcal{F})$ splits:

$$C(\xi) \simeq C(\eta) \oplus C(\xi/\eta).$$

But $\eta \cong C_*(f|E_1; \mathcal{F}_1)$ since $E_1^Q \simeq E^Q$ for every closed subset Q of $P_1^+(\sigma)$ and $\xi/\eta \cong C_*(f|E_2; \mathcal{F}_2)$ by excision of E_1. ($E_1|\sigma$ is a closed subset of every $E^Q|\sigma$ when Q contains $P_1^+(\sigma)$.)

If there is no fiberwise framed function we can first take the product with S^{2N} and divided all torsions by 2. □

As a consequence of this corollary we get a *gluing formula.* Suppose that E_1, E_2 are smooth bundles over B so that part of the boundary of E_1 agrees with part of the boundary of E_2:

$$\partial_0 E_1 = \partial_0 E_2.$$

Let $E_1 \cup E_2$ denote the union of E_1, E_2 along $\partial_0 E_1 = \partial_0 E_2$.

COROLLARY 3.17 (gluing formula).

$$\tau_k(E_1 \cup E_2; \mathcal{F}) = \tau_k(E_1; \mathcal{F}_1) + \tau_k(E_2; \mathcal{F}_2) - \tau_k(\partial_0 E_1; \mathcal{F}_0)$$

where $\mathcal{F}$ is a Hermitian coefficient system on $E_1 \cup E_2$ which restricts to $\mathcal{F}_1, \mathcal{F}_2, \mathcal{F}_0$ and where we assume that the terms are all defined.

PROOF. Choose a collar neighborhood D of $\partial_0 E_1$ in $E_1 \cup E_2$. Then D is a disk bundle over $\partial_0 E_1$ so, by the positive suspension theorem (3.2),

$$\tau_k(D; \mathcal{F}|D) = \tau_k(\partial_0 E_1; \mathcal{F}_0),$$

assuming this torsion is defined.

The union $E_1 \cup E_2$ is given by stacking $E_1 \coprod E_2$ on top of D. Therefore, by the stacking lemma, we have

$$\begin{aligned}\tau_k(E_1 \cup E_2; \mathcal{F}) &= \tau_k(D; \mathcal{F}|D) + \tau_k(E_1 \coprod E_2; \mathcal{F}) \\ &= \tau_k(\partial_0 E_1; \mathcal{F}_0) + \tau_k(E_1; \mathcal{F}_1) + \tau_k(E_2; \mathcal{F}_2),\end{aligned}$$

where we use the obvious torsion of disjoint union formula (3.12). □

Another example of the same idea is the decomposition of the higher torsion of a manifold bundle into the torsion of its boundary and the relative torsion.

COROLLARY 3.18 (relative torsion). Suppose that $E \to B$ is a smooth manifold bundle whose fiber M has boundary $\partial M = \partial_0 M \coprod \partial_1 M$ where each $\partial_i M$ is the fiber of a subbundle $\partial_i E$ of E. Then

$$\tau_k(E; \mathcal{F}) = \tau_k(E, \partial_0 E; \mathcal{F}) + \tau_k(\partial_0 E; \mathcal{F})$$

when the terms are defined.

PROOF. This follows from the gluing formula since E is equivalent to the union of $\partial_0 E \times I$ and E along their common boundary $d_0 E \times 0$. □

One final example of the same idea shows why the Suspension Theorem is "geometrically obvious."

COROLLARY 3.19 (easy geometric suspension). Suppose that $E \to B$ is a bundle with closed manifold fibers and $\mathcal{F}$ is a Hermitian coefficient system on E. Then

$$\tau_k(E \times D^m, E \times S^{m-1}; \pi^* \mathcal{F}) = (-1)^m \tau_k(E; \mathcal{F})$$

assuming the terms are defined.

PROOF. The theorem is a tautology for $m = 0$. So suppose by induction that the theorem holds for $m \geq 0$. Then, writing $E \times S^m$ as a union of $E \times D^m_-$ and $E \times D^m_+$ along $E \times S^{m-1}$, the stacking lemma gives:

$$\begin{aligned}\tau_k(E \times S^m) &= \tau_k(E \times D^m_-) + \tau_k(E \times D^m_+, E \times S^{m-1}) \\ &= \tau_k(E) + (-1)^m \tau_k(E)\end{aligned}$$

by induction and Theorem 2.34. So, the relative torsion formula above gives

$$\begin{aligned}\tau_k(E \times D^{m+1}, E \times S^m) &= \tau_k(E \times D^{m+1}) - \tau_k(E \times S^m) \\ &= [1 - (1 + (-1)^m)]\tau_k(E) = (-1)^{m+1}\tau_k(E).\end{aligned}$$

□

Suppose that $\Sigma(f)$ is a disjoint union of topologically closed subsets

$$\Sigma(f) = X_1 \coprod X_2 \coprod \cdots \coprod X_m$$

so that $X[i] = X_1 \coprod \cdots \coprod X_i$ is poset closed in $\Sigma(f)$ over every simplex in B, i.e., so that there is no trajectory of ∇f going from any point in X_i to any point in X_j for $j > i$. Then there are corresponding subfunctors $\xi[i]$ of $\xi = C_*(f; \mathcal{F})$. If the subquotients $\xi_i = \xi[i]/\xi[i-1]$ are acyclic then, by the Splitting Lemma the cellular chain complex functor of f is homotopic to a disjoint union of these subquotients:

$$C_*(f;\mathcal{F}) = \xi \simeq \xi_1 \oplus \cdots \oplus \xi_m.$$

We need to know that the algebraic subquotients, ξ_i, are determined by the geometry of the trajectories of ∇f coming out of the critical points in X_i. We call this the local equivalence lemma.

3.5. Local equivalence lemma

We state the lemma first and give the definitions afterwards. The *local equivalence lemma* assumes that we have two fiberwise oriented GMF's $f : E \to \mathbb{R}$, $f' : E' \to \mathbb{R}$ which are *geometrically equivalent* on "convex" subsets $S \subseteq E$ and $S' \subseteq E'$ in the sense that there is a diffeomorphism $\psi : S \to S'$ over B which is an isometry on each fiber covered by an isomorphism of coefficient sheaves $\widetilde{\psi} : \mathcal{F} \to \mathcal{F}'$ so that $f'\psi = f|S + c_t$ where c_t is a function of $t = p(x) \in B$. In other words, the following diagram commutes.

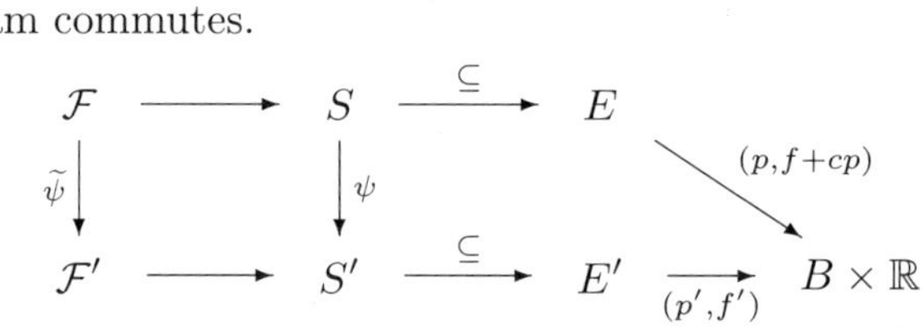

Note that, since ψ is a fiberwise isometry, its derivative takes ∇f to $\nabla f'$.

LEMMA 3.20 (local equivalence lemma). *Under these conditions, the local subquotient of $C_*(f;\mathcal{F})$ corresponding to S is isomorphic to the portion of $C_*(f';\mathcal{F}')$ corresponding to S'. In particular, the cellular chain complex functor of S is well defined without reference to the ambient bundle E.*

The idea and basic properties of convex sets comes from discussions with Gabriel Katz.

DEFINITION 3.21 (convex set). Let T be a compact subset of B. Then a compact subset S of $E|T$ will be called *convex* over T (with respect to $f : E \to \mathbb{R}$ and the vertical metric) if there are functions $a, b : T \to \mathbb{R}$, written $t \mapsto a_t, b_t$ with $a_t < b_t$ for all $t \in T$, satisfying the following.

(1) The fiber S_t of S over $t \in T$ is a codimension 0 submanifold of M_t (with corners) which varies smoothly with $t \in T$. (See Figure 1.)
(2) All critical points y of f_t in S_t have critical value
$$a_t < f_t(y) < b_t.$$
(3) The vertical gradient of f is nowhere zero on the vertical boundary of S, i.e., $\nabla f_t \neq 0$ along $\partial(S_t)$ for each $t \in T$.
(4) There are compact subsets A_t, B_t of $\partial(S_t)$ which vary smoothly with $t \in T$ satisfying the following.
   (a) $f_t(x) \leq a_t$ for all $x \in A_t$.
   (b) $f_t(x) \geq b_t$ for all $x \in B_t$.
   (c) Every point in $\partial(S_t)$ which is incident under a critical point of f_t in S_t lies in A_t and ∇f_t points inward at that point (so that the smooth flow of ∇f_t carries A_t into the interior of S_t).

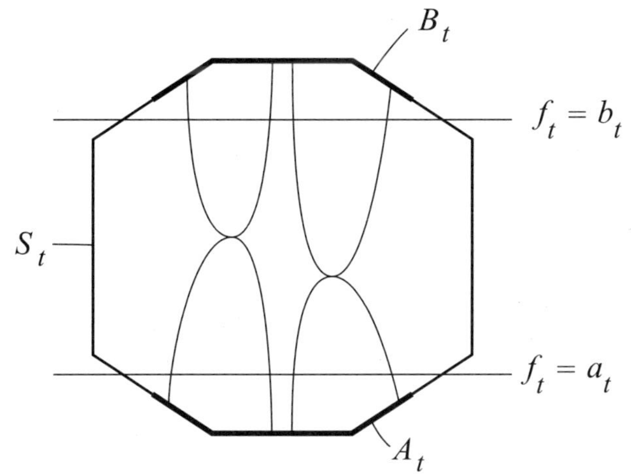

FIGURE 1. S_t (octagon) is called *convex* if A_t, B_t, a_t, b_t exist.

(d) Every point in $\partial(S_t)$ which is incident over a critical point of f_t in S_t lies in B_t and ∇f_t points outward at that point (so that the smooth flow of $-\nabla f_t$ carries B_t into the interior of S_t).

One easy example of a convex set is given by

$$S_t = f_t^{-1}[a_t, b_t]$$

where $a_t < b_t$ are regular values of f_t which vary smoothly with t. Suppose also that $\partial_0 M \subseteq f^{-1}(-\infty, a_t)$ and $\partial_1 M_t \subseteq f^{-1}(b_t, \infty)$. Then we can take $A_t = f_t^{-1}(a_t)$ and $B_t = f_t^{-1}(b_t)$.

Another example is given, in two steps, as follows. First, let S_t be a cobordism from A_t to B_t where these are codimension zero submanifolds of $f_t^{-1}(a_t), f_t^{-1}(b_t)$ respectively where $a_t < b_t$ are regular values of f_t and

$$\partial S_t = A_t \cup B_t \cup \partial A_t \times I.$$

Here, the side set $\partial A_t \times I$ should be a union of trajectories of ∇f_t going from ∂A_t up to $\partial B_t \cong \partial A_t$. There is a problem with this example. Along the corner set $\partial A_t \times \partial I$, the gradient of f_t is tangent to part of the boundary. So, S_t is not convex with this choice of A_t, B_t. But, if we delete a small collar neighborhood of the boundary of both A_t and B_t, the same set S_t will be convex.

The main nontrivial example is the following.

EXAMPLE 3.22 (main example). We start with a smooth bundle $E \to B$ with fiber Y. Suppose that $g : E \to \mathbb{R}$ is a fiberwise oriented GMF with the property that $\partial_0 E \subset g^{-1}(-\infty, 0)$, $\partial_1 E \subseteq g^{-1}(3\delta, \infty)$ and, for each $t \in B$, the critical values of g_t in the closed interval $[0, 3\delta]$ lie in the open interval $(\delta, 2\delta)$ for some $\delta > 0$. Let

$$E[0, 3\delta] = g^{-1}[0, 3\delta]$$

with fibers $Y_t[0, 3\delta] = g_t^{-1}[0, 3\delta]$ (and, more generally, $Y_t(J) = g_t^{-1}(J)$ for any $J \subseteq \mathbb{R}$).

Let γ, γ' be vector bundles over B of dimension m, m' with structure groups $O(m)$, $O(m')$ and let ϵD, $\epsilon D'$ be the ϵ–disk bundles of γ, γ' where ϵ is a positive real number so that $\epsilon^2 \geq 2\delta$.

Let S be the fiber product of $E[0,3\delta]$, ϵD and $\epsilon D'$ over B with the fiberwise product metric. Thus each fiber is a product

$$S_t = Y_t[0,3\delta] \times \epsilon D_t \times \epsilon D'_t$$

with the product metric. Let $f : S \to \mathbb{R}$ be given on each fiber S_t by

$$f_t(y,x,z) = g_t(y) - \|x\|^2 + \|z\|^2.$$

Then

$$\nabla f_t(y,x,z) = (\nabla g_t(y), -2x, 2z).$$

Consequently, the critical points of f_t are the same as those of $g_t|Y_t[0,3\delta]$.

$$\Sigma(f_t) = \Sigma(g_t|Y_t[0,3\delta]) \times 0 \times 0.$$

Then we claim that S is a convex set with respect to f where

(1) $T = B$.
(2) $a_t = 0$.
(3) $b_t = 3\delta$.
(4) $A_t = Y_t(0) \times \epsilon D_t \times 0 \ \cup \ Y_t[0,2\delta] \times \partial(\epsilon D_t) \times 0$.
(5) $B_t = Y_t(3\delta) \times 0 \times \epsilon D'_t \ \cup \ Y_t[\delta,3\delta] \times 0 \times \partial(\epsilon D'_t)$.

The definition is satisfied since all trajectories of ∇f_t emanating from critical points of f_t lie in $Y_t[\delta,3\delta] \times 0 \times \epsilon D'_t$ and the intersection of this set with ∂S_t is exactly B_t. Similarly, trajectories of ∇f_t going towards $\Sigma(f_t)$ lie in $Y_t[0,2\delta] \times \epsilon D_t \times 0$ which meets ∂S_t along A_t. The inequality $2\delta \leq \epsilon^2$ insures that $f_t \leq 0$ on $Y_t[0,2\delta] \times \partial(\epsilon D_t) \times 0 \subseteq A_t$ and $f_t \geq 3\delta$ on $Y_t[\delta,3\delta] \times 0 \times \partial(\epsilon D'_t) \subseteq B_t$.

Sometimes we need to deform the function f_t in the main example. This happens, for example, when we need f_t to be fiberwise framed.

EXAMPLE 3.23 (deformation of main example). Take $S_t = Y_t[0,3\delta] \times \epsilon D_t \times \epsilon D'_t$ and $f_t(y,x,z) = g_t(y) - \|x\|^2 + \|z\|^2$ as in the main example. Note that

$$\Sigma(f_t) \subseteq Y_t(\delta,2\delta) \times 0 \times 0.$$

Suppose that $\widetilde{f}_t$ is a deformation of f_t with support in a small neighborhood of $\Sigma(f_t)$, say

$$Y_t[\delta,2\delta] \times \beta D_t \times \beta D'_t \tag{3.1}$$

for some small $\beta > 0$. If

$$2\beta^2 < \delta$$

then we claim that S will be a convex set for the new function $\widetilde{f}$ where

(1) $T = B$.
(2) $a_t = \delta/2$.
(3) $b_t = 5\delta/2$.
(4) $A_t = Y_t(0) \times \epsilon D_t \times \beta D'_t \ \cup \ Y_t[0,2\delta] \times \partial(\epsilon D_t) \times \beta D'_t$.
(5) $B_t = Y_t(3\delta) \times \beta D_t \times \epsilon D'_t \ \cup \ Y_t[\delta,3\delta] \times \beta D_t \times \partial(\epsilon D'_t)$.

Any trajectory of $\nabla\widetilde{f}_t$, before reaching ∂S_t must pass through the boundary of the set (3.1). At that point it becomes equal to ∇f_t and thus goes up to B_t as does any point in the set $Y_t[\delta,3\delta] \times \beta D_t \times \epsilon D'_t$. Since the x–coordinate is now allowed to be in βD_t we need to subtract β^2 from b_t.

Another example that we will use is the following.

EXAMPLE 3.24 (second example). Let $g : E \to \mathbb{R}$, $E[0, 3\delta]$ with fibers $Y_t[0, 3\delta]$ be as in the main example. Also, let γ, γ' be as before and let ϵD, $\epsilon D'$ be the ϵ–disk bundles of γ, γ' where ϵ is a positive real number so that

$$3\delta \leq \epsilon^3.$$

As before, let S be the fiber product of $E[0, 3\delta]$, ϵD and $\epsilon D'$ over B with the fiberwise product metric. Take the product with J^2 where $J = [-\epsilon, \epsilon]$. Then $W = S \times J^2$ is a bundle over $B \times J$ whose fiber over $(t, u) \in B \times J$ is

$$W_{t,u} = S_t \times J = Y_t[0, 3\delta] \times \epsilon D_t \times \epsilon D'_t \times J.$$

We take the fiberwise product metric on W. Let $f : W \to \mathbb{R}$ be given on each fiber $W_{t,u}$ by

$$f_{t,u}(y, x, z, w) = g_t(y) - \|x\|^2 + \|z\|^2 + w^3 - uw.$$

Then

$$\nabla f_{t,u}(y, x, z, w) = (\nabla g_t(y), -2x, 2z, 3w^2 - u).$$

So,

$$\Sigma(f_{t,u}) = \Sigma(g_t | Y_t[0, 3\delta]) \times 0 \times 0 \times \pm\sqrt{u/3}.$$

The function f has a problem. It may have a singularity of the form $y^3 + w^3$. These bad singularities occur when x, z, w, u are all zero and x is a b-d singularity of g_t. To fix this we choose $\beta > 0$ so that $\beta < 2\epsilon/3$ and $2\beta^2 < \delta$ and note that the bad set lies in the interior of the set

$$Y[\delta, 2\delta] \times \beta D \times \beta D' \times [-\beta, \beta] \times B \times [-\tfrac{2\epsilon}{3}, \tfrac{2\epsilon}{3}]. \tag{3.2}$$

Suppose that h is a deformation of f with support in the interior of this set so that h is a fiberwise oriented GMF. (For example, if g is fiberwise framed then so is f outside of the set (3.2) so we can use the C^1–local framed function theorem (2.5) to find h.)

We claim that W is a convex set with respect to the new function h where

(1) $T = B \times J$.
(2) $a_{t,u} = \delta/2$.
(3) $b_{t,u} = 5\delta/2$.
(4) $A_{t,u} = \partial S_t \cap \big(Y_t[0, 2\delta] \times \epsilon D_t \times \beta D'_t \times [-\epsilon, \tfrac{2\epsilon}{3}]\big)$.
(5) $B_{t,u} = \partial S_t \cap \big(Y_t[\delta, 3\delta] \times \beta D_t \times \epsilon D'_t \times [-\tfrac{2\epsilon}{3}, \epsilon]\big)$.

The definition is satisfied since all trajectories of $\nabla h_{t,u}$ emanating from critical points of $h_{t,u}$ lie in $Y_t[\delta, 3\delta] \times \beta D_t \times \epsilon D'_t \times [-\frac{2\epsilon}{3}, \epsilon]$ and similarly, those going towards $\Sigma(h_{t,u})$ lie in $Y_t[0, 2\delta] \times \epsilon D_t \times \beta D'_t \times [-\epsilon, \frac{2\epsilon}{3}]$.

Before proving the local equivalence lemma we make some basic observations about convex sets. A convex set S will contain all *multiple trajectories* of ∇f_t (i.e., a sequence of trajectories so that the target of each is the source of the next) passing between two critical points in S_t. The reason is that if a trajectory of ∇f_t coming from a critical point of f_t in S_t passes out of S_t then it goes through a point in ∂S_t where ∇f_t points outward (away from S_t) and $f_t \geq b_t$. It has no chance to arrive later at a point in ∂S_t where ∇f_t points inward and $f_t \leq a_t < b_t$.

Suppose that $T = \sigma$ is a small simplex in B. Then this property of convex sets implies that the subset Q of the poset $P(\sigma)$ given by the critical points of $f|\sigma$ in $S|\sigma$ is *half closed* (the intersection of an open and closed subset) provided that σ is sufficiently small.

Combinatorially, a subset Q of a poset P is half closed if and only if $x < y < z$ in P with $x, z \in Q$ implies that $y \in Q$. Since we are using the minimal partial ordering, the critical points of f_t in S_t for t fixed satisfies this property as we explained above. However, for critical points in $Q \times \sigma \subseteq \Sigma(f)|\sigma$ it could happen that

$$(x, t_1) < (y, t_1) \quad \text{and} \quad (y, t_2) < (z, t_2)$$

for two different points $t_1, t_2 \in \sigma$. But this would imply that $f_{t_1}(y) > b_t$ and $f_{t_2}(y) < a_t$ which will not happen if σ is chosen sufficiently small.

DEFINITION 3.25 (local subquotient). A *local subquotient* of an expansion functor

$$\xi : \operatorname{simp} X \to \mathcal{W}h_\bullet(R, G)$$

is defined to be the expansion functor generated by a *family of half closed subsets*. These are given by $Q(x) \subseteq P(x)$ for each $(x, [k])$ in $\operatorname{simp} X$ so that the structure morphisms

$$(P(y), \phi(y)) \to (P(x), \phi(x))|y$$

send $Q(y)$ onto $Q(x)$ and so that, if one member of a collapsing pair lies in $Q(y)$, so does the other. (Recall that $P(y)$ is a disjoint union of collapsing pairs x_i^+, x_i^- and another set A which maps bijectively onto $P(x)$.)

The purpose of local subquotients is to decompose the expansion functor ξ as illustrated by the following proposition.

PROPOSITION 3.26 (acyclic local subquotients). Suppose that ξ is an (R, G)–expansion functor as above and $Q_1, \cdots Q_m$ are families of half closed subsets of P so that

$$P(x) = \coprod Q_i(x)$$

for every $(x, [k]) \in \operatorname{simp} X$. Suppose further that the local subquotient η_i generated by Q_i is acyclic for every i. Then ξ is homotopic to the direct sum of functors

$$\xi \simeq \bigoplus \eta_i : \operatorname{simp} X \to \mathcal{W}h(R, G).$$

PROOF. Use the local splitting lemma (3.13). Over each simplex $(x, [k])$ in X the Δ^k–family of chain complexes $\xi(x)$ has a filtration with acyclic subquotients $\eta_i(x)$. Therefore, by induction on k, we can eliminate (make 0) all cross terms in the twisting cochain $\phi(x)$. This defines a simplicial homotopy $\phi \simeq \bigoplus \phi_i$ where ϕ_i is the twisting cochain of Q_i. But we also have a monomial equivalence

$$\bigoplus \eta_i(x) = (\vee Q_i(x), \bigoplus \phi_i(x)) \to (P(x), \bigoplus \phi_i(x)).$$

□

PROOF OF LOCAL EQUIVALENCE LEMMA. By making a_t slightly larger and b_t slightly smaller we may assume that $f_t < a_t$ on A_t and $f_t > b_t$ on B_t. Then the smooth flow generated by ∇f_t translates A_t and B_t into compact subsets A_t' and B_t' of the level surfaces $X_t = f_t^{-1}(a_t)$ and $Y_t = f_t^{-1}(b_t)$. Since ∇f_t points into the interior of S_t along A_t, the set A_t' will lie in the interior of S_t as will the set B_t'.

Let V be the union over all $t \in B$ of the set V_t of all points $x \in S_t$ so that

$$a_t \leq f_t(x) \leq b_t.$$

Let $Z = \cup_{t \in B} Z_t$ where Z_t is the set of points in V_t which are incident over $\Sigma(f_t|S_t)$. Then V, Z are compact and

$$Z_t \cap Y_t \subseteq B'_t \subseteq Y_t \cap \text{int} S_t.$$

Note that V, Z are intrinsic to S, i.e., independent of the embedding $S \subseteq E$. So we can use them to construct the intrinsic subquotient functor given by S.

For every sufficiently small transverse simplex σ in B let $Q^+(\sigma)$ be the half-closed subset of $P^+(\sigma)$ corresponding to the singularities of $f|\sigma$ in $S|\sigma$. ($Q^+(\sigma)$ is the set of components of $\Sigma(f|\sigma) \cap S$.) For every closed subset L of $Q^+(\sigma)$ let $V^L|\sigma$ be the subset of $V|\sigma$ consisting of all points which are not incident over $Q^+(\sigma)/L$. In particular, $V^\emptyset(\sigma) = (S - Z)|\sigma$. Take the total singular complex of the pair $(V|\sigma, V^\emptyset|\sigma)$ together with the filtration given by the subsets $V^L|\tau$ for all faces τ of σ. We claim that this gives a filtered chain complex which is filtered quasi-isomorphic to the total complex of the subquotient functor.

We need to compare this intrinsically defined filtered complex with the actual subquotient functor. Let $K^+(\sigma)$ be the closure in $P^+(\sigma)$ of the set of all components x containing elements with $f_t(x) < a_t$. We also add to $K^+(\sigma)$ members of collapsing and expanded pairs whose mate is already in $K^+(\sigma)$. Then $K^+(\sigma)$ will be disjoint from $Q^+(\sigma)$ and $K^+(\sigma) \coprod Q^+(\sigma)$ will be closed. We also know, by Proposition 2.6, that $f_t(x) < b_t$ for every $t \in \sigma$ and every $x \in \Sigma(f_t)$ in any component in $K^+(\sigma)$. By definition of a convex set we also know that the set of all points in $S|\sigma$ which are incident over a point in the complement of $Q^+(\sigma) \cup K^+(\sigma)$ is a closed subset disjoint from Z_t.

Given any closed subset L of $Q^+(\sigma)$ and any face τ of σ, the inclusion maps

$$(V^L|\tau, V^\emptyset|\tau) \to (E^{L \cup K^+(\sigma)} \cap W|\tau, E^{K^+(\sigma)} \cap W|\tau) \xrightarrow{\simeq} (E^{L \cup K^+(\sigma)}|\tau, E^{K^+(\sigma)}|\tau)$$

induce isomorphisms in homology. Here W is given by

$$W_t = f_t^{-1}(-\infty, b_t]$$

and the deformation retraction onto the intersection with W is given by the smooth flow of $-\nabla f$. The first mapping is excisive.

By Theorem 2.14, the twisting cochain generated by $Q^+(\sigma)$ is intrinsically defined. So the subquotient functor is isomorphic to the one for S'. □

If we take the construction of the intrinsically defined subquotient functor for S applied to the main example (3.22), we get an m–fold suspension of the expansion functor of f by the Thom isomorphism theorem (as in the case of the negative suspension lemma 3.9).

PROPOSITION 3.27 (subquotient functor of main example). *The intrinsic cellular chain complex functor for the convex set in Example 3.22 is the m–fold suspension of the expansion functor for $f : E \to I$:*

$$C_*(h; \rho) = \Sigma^m C_*(f; \rho).$$

Now we discuss the second example (3.24). We will not obtain a complete description of the subquotient functor because there is a problem with the b-d points. The problem is that

$$h(x, y) = x^3 + y^3$$

is not an allowable singularity. Suspensions of these singularities occur in the second example at the points where both f and g have b-d points.

3.6. Product formula

We are now ready to prove one of the main formulas.

LEMMA 3.28 (product formula). *Suppose Y is a compact smooth manifold so that $\partial_0 E \times \partial Y = \emptyset$, i.e., either Y is closed or $\partial_0 E$ is empty. Then,*

$$\tau_k(E \times Y, \partial_0 E \times Y; \mathcal{F}') = \chi(Y)\tau_k(E, \partial_0 E; \mathcal{F})$$

assuming that $\tau_k(E, \partial_0 E; \mathcal{F})$ is defined. Here $\mathcal{F}'$ is the pullback of $\mathcal{F}$ along the projection $E \times Y \to E$.

REMARK 3.29 (definition of higher torsion). This is the key lemma that we are using in this paper to circumvent one of the difficulties in the definition of higher FR–torsion in the relative case. The proof is the same as in [**Igu02**] except that we must be careful to avoid circular reasoning.

PROOF. The assumption $\partial_0 E \times \partial Y = \emptyset$ implies that the boundary of $E \times Y$ is a disjoint union

$$\partial(E \times Y) = \partial_0 E \times Y \coprod (\partial_1 E \times Y \cup E \times \partial Y).$$

Choose a Morse function $g : Y \to [0, n+1]$ so that $g^{-1}(n+1) = \partial Y$ with critical points $y_0, y_1, \cdots, y_n$ having distinct integer critical values $g(y_i) = i$. Next, suppose there exists a fiberwise framed function

$$f : (E, \partial_0 E) \to (I, 0)$$

so that $f^{-1}(1) = \partial_1 E$. Consider the smooth function

$$h : E \times Y \to [0, n+2]$$

given by $h(x, y) = f(x) + g(y)$.

For each $t \in B$ we get a smooth function $h_t : M_t \times Y \to \mathbb{R}$ with critical set $\Sigma(h_t) = \Sigma(f_t) \times \Sigma(g)$. At each critical point (x, y_i) we can rearrange the local coordinates so that the negative quadratic terms for y_i go to the beginning. This gives h the structure of a fiberwise framed function.

Since the gradient of h_t points inward along $\partial_0 M_t \times Y$ and outward along

$$\partial_1 M_t \times Y \cup M_t \times \partial Y$$

for all $t \in B$ we can modify h in a small neighborhood of $\partial(E \times Y)$ without introducing new critical points so that

$$h^{-1}(0) = \partial_0 E \times Y, \quad \partial_1 E \times Y \cup E \times \partial Y \subseteq h^{-1}[n+1, n+2].$$

Then, $h^{-1}[0, i+1]$, $i = 0, \cdots, n$ gives a filtration of $E \times Y$ resulting in a filtration

$$\xi_0 \subseteq \xi_1 \subseteq \cdots \subseteq \xi_n = \xi$$

of the family of chain complexes $\xi = C_*(h; \mathcal{F}')$. Since y_0 has index 0, $\xi_0 \cong C_*(f)$ and, for each i, the quotient ξ_i/ξ_{i-1} is isomorphic to a suspension of $C_*(f; \mathcal{F})$

$$\xi_i/\xi_{i-1} \cong \Sigma^{\mathrm{ind}(y_i)} C_*(f; \mathcal{F}).$$

By the Suspension Theorem this has torsion

$$\tau_k(C(\xi_i/\xi_{i-1})) = (-1)^{\mathrm{ind}(y_i)} \tau_k(CC_*(f; \mathcal{F})).$$

The lemma now follows from the Splitting Lemma

$$\begin{aligned}\tau_k(E \times Y, \partial_0 E \times Y; \mathcal{F}') &= \tau_k(C(\xi)) = \sum \tau_k(C(\xi_i/\xi_{i-1})) \\ &= \sum (-1)^{\operatorname{ind}(y_i)} \tau_k(CC_*(f; \mathcal{F})) = \chi(Y)\tau_k(E, \partial_0 E; \mathcal{F}).\end{aligned}$$

Now suppose that $\dim M < \dim B$ and there is no fiberwise framed function for $(E, \partial_0 E)$. To avoid circular reasoning we cannot use the usual trick of taking the product with an even dimensional sphere. Instead we take the product with two spheres $S^{2N}, S^{2N'}$ with N, N' very large.

Since $E \times S^{2N}$, $E \times S^{2N'}$ have large dimensional fibers, the lemma holds for them. Thus:

$$\begin{aligned}\tau_k(E \times S^{2N}, \partial_0 E \times S^{2N}; \mathcal{F}') &= \frac{1}{2}\tau_k(E \times S^{2N} \times S^{2N'}, \partial_0 E \times S^{2N} \times S^{2N'}; \mathcal{F}''') \\ &= \tau_k(E \times S^{2N'}, \partial_0 E \times S^{2N'}; \mathcal{F}'').\end{aligned}$$

Therefore, $\tau_k(E \times S^{2N}, \partial_0 E \times S^{2N}; \mathcal{F}')$ is *independent of the choice of* N as long as N is sufficiently large. So, we can *define* $\tau_k(E, \partial_0 E; \mathcal{F})$ to be half of this:

$$\tau_k(E, \partial_0 E; \mathcal{F}) := \frac{1}{2}\tau_k(E \times S^{2N}, \partial_0 E \times S^{2N}; \mathcal{F}'). \tag{3.3}$$

The above argument shows that this equation holds whenever $(E, \partial_0 E)$ admits a fiberwise framed function.

Now we can complete the proof of the lemma (suppressing the coefficients and the $\partial_0 E$ terms from the notation).

$$\tau_k(E \times Y) = \frac{1}{2}\tau_k(E \times Y \times S^{2N}) = \frac{1}{2}\chi(Y)\tau_k(E \times S^{2N}) = \chi(Y)\tau_k(E).$$

□

3.7. Transfer for coverings

A Morse function on a manifold M gives a Morse function on any covering of M and the associated chain complexes are related by a simple transfer formula. We use this elementary observation to explain why the Riemann zeta function appears in the formula for the higher FR–torsion invariant.

This argument is a simplification of an argument which appears in [**Igu02**]. I should thank William Hoffman and Neil Stoltzfus for helping me to clarify this idea.

Suppose that G is a finite group which acts freely on the total space E so that the action is smooth and commutes with the projection $p : E \to B$. Then we get an induced bundle

$$E_G = E/G \to B$$

with fiber $M_G = M/G$. Note that E is a finite covering space of E_G. If H is any subgroup of G, then $E_H = E/H$ is also a finite covering space of E_G and we have covering maps

$$E \to E_H \to E_G.$$

Conversely, any finite connected covering $\widetilde{X} \to X$ has this form where G is the quotient of $\pi_1 X$ by the largest normal subgroup contained in the image of $\pi_1 \widetilde{X}$.

THEOREM 3.30 (transfer for coverings of E). *Suppose that V is a unitary representation of H so that $\pi_1 B$ acts trivially on $H_*(M_H;V) \cong H_*(M_G;\operatorname{Ind}_H^G V)$. Then*

$$\tau_k(E_G;\operatorname{Ind}_H^G V) = \tau_k(E_H;V).$$

PROOF. Suppose first that there is a fiberwise framed function $f_G : E_G \to \mathbb{R}$. Composing with the covering maps $E \to E_H \to E_G$ we get induced fiberwise framed functions $f_H : E_H \to \mathbb{R}$ and $f : E \to \mathbb{R}$. The cellular chain complex $C_*(f)$ is a family of free $\mathbb{C}G$–complexes over B and the higher torsion invariants $\tau_k(E_H;V)$ are induced by

$$\begin{aligned} C_*(f_H;V) &= C_*(f) \otimes_{\mathbb{C}H} V \\ &\cong C_*(f) \otimes_{\mathbb{C}G} \mathbb{C}G \otimes_{\mathbb{C}H} V \\ &= C_*(f) \otimes_{\mathbb{C}G} \operatorname{Ind}_H^G V = C_*(f_G;\operatorname{Ind}_H^G V) \end{aligned}$$

considered as a mapping from B into the space of acyclic chain complexes (using the canonical cone construction). Consequently, they have the same torsion:

$$\tau_k(E_G;\operatorname{Ind}_H^G V) = \tau_k(C_*(f_G;\operatorname{Ind}_H^G V)) = \tau_k(C_*(f_H;V)) = \tau_k(E_H;V).$$

If the dimension of M is not large enough to admit a fiberwise framed function we first take the product of all bundles with a large even dimensional sphere S^{2N}. Then there is a fiberwise framed function on $E_G \times S^{2N}$ and the above argument tells us that

$$\tau_k(E_G \times S^{2N};\operatorname{Ind}_H^G V) = \tau_k(E_H \times S^{2N};V).$$

Divide by 2 to get the theorem. □

Now we specialize to cyclic groups. Let $Z_n = \mathbb{Z}/n$ and for z any n–th root of unity let

$$\rho_z : Z_n \to U(1)$$

be the 1–dimensional representation of Z_n which sends the generator of Z_n to z. Then, by Frobenius reciprocity, we have

$$\operatorname{Ind}_{Z_n}^{Z_{nm}}(\rho_z) = \bigoplus_{\rho|H=\rho_z} \rho = \bigoplus_{\zeta^m=z} \rho_\zeta$$

Consequently we have:

COROLLARY 3.31 (transfer for cyclic coverings).

$$\tau_k(E/Z_n;\rho_z) = \sum_{\zeta^m=z} \tau_k(E/Z_{nm};\rho_\zeta).$$

Now consider the special case where $M = S^1 = U(1)$ and $E \to B$ is the universal S^1–bundle

$$EU(1) \to BU(1) = \mathbb{C}P^\infty.$$

Then $H^*(BU(1)) = \mathbb{Z}[c]$ and the circle bundle $EU(1)/Z_m \to BU(1)$ is classified by a mapping

$$\phi_m : BU(1) \to BU(1)$$

satisfying $\phi_m^*(c) = mc$ ($\Rightarrow \phi_m^*(c^k) = m^k c^k$).

More generally, there is a morphism of S^1–bundles:

$$\begin{array}{ccc} EU(1)/Z_{nm} & \longrightarrow & EU(1)/Z_n \\ \downarrow & & \downarrow \\ BU(1) & \xrightarrow[\phi_m]{} & BU(1) \end{array}$$

Consequently, by naturality of the higher FR–torsion invariant, we have

$$\tau_k(EU(1)/Z_{nm}; \rho_z) = m^k \tau_k(EU(1)/Z_n; \rho_z)$$

for any n–th root of unity z. Since this holds in the universal case, it holds for all oriented S^1–bundles. Combining this with the second transfer formula (3.31) we get:

$$\tau_k(E/Z_s; \rho_z) = m^k \sum_{\zeta^m = z} \tau_k(E/Z_s; \rho_\zeta).$$

Here $s = nm$. Now we use the elementary fact that the polylogarithm function

$$L_{k+1}(z) = \Re\left(\frac{1}{i^k}\sum_{n=1}^{\infty}\frac{z^n}{n^{k+1}}\right) \tag{3.4}$$

is, up to a constant multiple, the only smooth function on $z \in U(1) - \{1\}$ satisfying the equation

$$L_{k+1}(z) = m^k \sum_{\zeta^m = z} L_{k+1}(\zeta)$$

for all integers $m \neq 0$. (See [**Mil83**] for a proof of the stronger statement that these are the only *continuous* functions on $U(1) - \{1\}$ satisfying (3.4).) The conclusion is that

$$\tau_k(EU(1)/Z_n; \rho_z) = n^k a_k L_{k+1}(z) c^k \tag{3.5}$$

where a_k is a real number depends only on k. See [**Igu02**] for the missing steps in this argument and for the proof that:

$$a_k = -\frac{1}{k!} \tag{3.6}$$

for all k.

THEOREM 3.32 (torsion of circle bundles). *Let $S^1(\lambda)$ be the circle bundle associated to the complex line bundle λ over B. Then*

$$\tau_k(S^1(\lambda)/Z_n; \rho_z) = -\frac{n^k}{k!} L_{k+1}(z) c_1(\lambda)^k.$$

Taking the special case $n = 1 = z$ we get the higher torsion of the universal oriented S^1–bundle:

$$\tau_{2k}(EU(1)) = -\frac{1}{(2k)!} L_{2k+1}(1) c^{2k} = (-1)^{k+1}\zeta(2k+1) ch_{2k}(\gamma) \tag{3.7}$$

where γ is the canonical complex line bundle over $BU(1)$.

COROLLARY 3.33 (torsion of lens space bundles). Let $S^{2n-1}(\xi) \to B$ be the sphere bundle associated to a complex n–plane bundle ξ. Then

$$\tau_k(S^{2n-1}(\xi)/Z_m; \rho_z) = -m^k L_{k+1}(z) ch_k(\xi)$$

where $L_{k+1}(z)$ is given in (3.4) above.

PROOF. This is Theorem 5.7.12 in [**Igu02**]. An outline of the proof goes as follows.

Using the splitting principle we may assume that ξ is a sum of line bundles $\xi = \oplus\lambda_i$. This implies that the sphere bundle $S^{2n-1}(\xi)$ is a fiberwise join of the circle bundles associated to the λ_i and

$$S^{2n-1}(\xi)/Z_m \cong *_i S^1(\lambda_i)/Z_m.$$

Consequently, by the Splitting Lemma, we have

$$C_*(S^{2n-1}(\xi)/Z_m) \cong \bigoplus_i C_*(S^1(\lambda_i)/Z_m).$$

So,

$$\begin{aligned}\tau_k(S^{2n-1}(\xi)/Z_m;\rho_z) &= \sum_{i=1}^{n} \tau_k(S^1(\lambda_i)/Z_m;\rho_z)\\ &= -\frac{m^k}{k!}L_{k+1}(z) = -m^k L_{k+1}(z)ch_k(\xi).\end{aligned}$$

□

3.8. More transfer formulas

Another way to do the transfer is on the base. If

$$\pi : \widetilde{B} \to B$$

is a finite covering map then we recall that the *transfer* or *push-down*

$$tr = tr_B^{\widetilde{B}} = \pi_* : H^q(\widetilde{B};A) \to H^q(B;A)$$

with any additive group coefficients A is defined at the chain level by

$$\langle tr(c_q),\sigma\rangle = \sum_{\widetilde{\sigma}} \langle c_q,\widetilde{\sigma}\rangle \tag{3.8}$$

for any q–cocycle c_q on $\widetilde{B}$ (with coefficients in A) and any singular q–simplex $\sigma : \Delta^q \to B$. The sum is over all lifting $\widetilde{\sigma}$ of σ to $\widetilde{B}$.

Suppose that $p : E \to \widetilde{B}$ is a smooth bundle with $\partial E = \partial_0 E \coprod \partial_1 E$ and $\mathcal{F}$ is a Hermitian coefficient system on E so that $H_*(M,\partial_0 M;\mathcal{F}) = 0$. Then the higher relative FR–torsion of $(E,\partial_0 E)$ as a bundle over both $\widetilde{B}$ and B are defined and related by transfer.

THEOREM 3.34 (transfer for coverings of B).

$$\tau_k(E,\partial_0 E;\mathcal{F})_B = tr_B^{\widetilde{B}}(\tau_k(E,\partial_0 E;\mathcal{F})_{\widetilde{B}}).$$

REMARK 3.35 (torsion commutes with transfer). At the cellular chain complex level this theorem says that the torsion of the transfer is the transfer of the torsion, i.e.,

$$\tau_k(tr_B^{\widetilde{B}}(\xi)) = tr_B^{\widetilde{B}}(\tau_k(\xi)) \tag{3.9}$$

where the *transfer*, $tr_B^{\widetilde{B}}(\xi)$, of an expansion functor ξ on $\widetilde{B}$ is the expansion functor on B given by

$$tr_B^{\widetilde{B}}(\xi)(\sigma) = \bigoplus_{\widetilde{\sigma}} \xi(\widetilde{\sigma}). \tag{3.10}$$

where the direct sum is over all lifting $\widetilde{\sigma}$ of σ to $\widetilde{B}$.

PROOF. The proof is the same as that of the additivity theorem. We choose a fiberwise framed function

$$f : (E, \partial_0 E) \to (I, 0),$$

taking the product of E with S^{2N} if necessary. The cellular chain complex $C_*(f; \mathcal{F})$ gives a $(M_r(\mathbb{C}), U(r))$–expansion functor

$$\xi : \operatorname{simp} \widetilde{B} \to \mathcal{W}h^h_\bullet(M_r(\mathbb{C}), U(r)).$$

For any small simplex σ in B, the inverse image of σ in E is a disjoint union,

$$E|\sigma = \coprod_{\widetilde{\sigma}} E|\widetilde{\sigma},$$

over all liftings $\widetilde{\sigma}$ of σ to $\widetilde{B}$. Consequently, using the same function f, the resulting expansion functor on $\widetilde{B}$ is the transfer of ξ as given in (3.10). Thus it suffices to show that transfer commutes with torsion (3.9).

Using the 2–index theorem (Theorem 2.23), we can deform ξ into a family of complex matrices η parametrized by $\widetilde{B}$. The homotopy $\xi \simeq \eta$ give a homotopy between the push-down expansion functors

$$\overline{\xi} = tr_B^{\widetilde{B}}(\xi) : \operatorname{simp} B \to \mathcal{W}h^h_\bullet(M_r(\mathbb{C}), U(r))$$

$$\overline{\eta} = tr_B^{\widetilde{B}}(\eta) : \operatorname{simp} B \to \mathcal{W}h^{h[0,1]}_\bullet(M_r(\mathbb{C}), U(r))$$

and it remains only to show that the higher torsion of $\overline{\eta}$ is the transfer of the higher torsion of η.

The higher FR–torsion of $\overline{\xi}$ is given by integrating the Kamber-Tondeur form for $\overline{\eta}$ plus polynomial correction terms. Both of these obviously satisfy the property that they are additive with respect to direct sum. (They are both integrals of traces of products of matrix 1–forms which preserve the direct sum structure of the input matrices $\overline{\eta}$.) In other words,

$$\langle KT_k(\overline{\eta}), \sigma \rangle = \sum_{\widetilde{\sigma}} \langle KT_k(\eta), \widetilde{\sigma} \rangle$$

where $KT_k(\overline{\eta})$ is the $2k$–cocycle on $\operatorname{simp} B$ which gives the higher FR–torsion $\overline{\eta}^*(\tau_k)$ of $(E, \partial_0 E; \mathcal{F})$ over B. Thus, $\tau_k(\overline{\eta}) = tr_B^{\widetilde{B}}(\tau_k(\eta))$ at the chain level. ☐

Now suppose that

$$f : E \to \mathbb{R}$$

is a *fiberwise framed Morse function*, i.e., a fiberwise framed function without b-d points. Then the singular set $\Sigma = \Sigma(f)$ is a disjoint union

$$\Sigma(f) = \coprod_{i=0}^{n} \Sigma^i(f)$$

where each $\Sigma^i(f)$ is a finite covering space of B.

Suppose we have a smooth bundle

$$Y \to D \to E$$

and $\mathcal{F}$ is a Hermitian coefficient system over D with respect to which Y is acyclic, i.e., $H_*(Y; \mathcal{F}) = 0$. We consider D as a bundle over B. This bundle has corners which we can ignore since we will be enclosing the singular set of a fiberwise framed function in a disjoint union of convex sets disjoint from the corner set.

The transfer theorem for coverings of B (Theorem 3.34) applies to the restriction of D to $\Sigma^i = \Sigma^i(f)$ to give

$$\tau_k(D|\Sigma^i;\mathcal{F})_B = tr_B^{\Sigma^i}(\tau_k(D|\Sigma^i;\mathcal{F})_{\Sigma^i}). \tag{3.11}$$

We claim that the alternating sum of these terms gives the higher torsion of D as a bundle over B. This is a special case of the main theorem of this paper.

THEOREM 3.36 (transfer without b-d points). *Given a fiberwise framed Morse function $f : E \to \mathbb{R}$, a smooth bundle $D \to E$ and Hermitian coefficient system $\mathcal{F}$ over D as above, we get the following.*

$$\tau_k(D;\mathcal{F})_B = \sum_{i=0}^{n}(-1)^i\tau_k(D|\Sigma^i;\mathcal{F})_B.$$

REMARK 3.37 (push-down without b-d points). By (3.11) this can be expressed as

$$\tau_k(D;\mathcal{F}) = p_*^{\Sigma}(\tau_k(D|\Sigma;\mathcal{F})_\Sigma)$$

where the *push-down operator*,

$$p_*^{\Sigma} : H^{2k}(\Sigma(f);\mathbb{R}) \to H^{2k}(B;\mathbb{R}),$$

is given by the alternating sum of transfer maps:

$$p_*^{\Sigma} := \sum_{i=0}^{n}(-1)^i tr_B^{\Sigma^i}.$$

PROOF. We will enclose the singular set of f in a subset T of E and find a fiberwise framed function h on D which is locally equivalent over T to the convex set in the main example (3.22). Then Proposition 3.27 will give us the expansion functor of h.

Let $T^i \subseteq E$ be a tubular neighborhood of $\Sigma^i = \Sigma^i(f)$ so that the fiber of the projection map

$$\pi : T^i \to \Sigma^i$$

is a product of two disks, an i–disk and a twisted $(n-i)$–disk.

$$T^i \cong \epsilon D^i \times \epsilon D^{n-i}(\gamma_i)$$

where ϵD^i is the disk of radius ϵ in $\mathbb{R}^i$:

$$\epsilon D^i = \{x \in \mathbb{R}^i \mid \|x\| \leq \epsilon\}$$

and $\epsilon D^{n-i}(\gamma_i)$ is the ϵ–disk bundle of an $(n-i)$–plane bundle γ_i over Σ^i. (γ_i is the positive eigenspace bundle of $D^2 f$.)

By the parametrized Morse lemma we may assume that $f|T^i$ is given by

$$f(x,z) = f(s) - \|x\|^2 + \|z\|^2 \tag{3.12}$$

where $s = \pi(z) \in \Sigma^i \subseteq E$ (assuming $\epsilon > 0$ is sufficiently small).

Since $\pi : T^i \to \Sigma^i$ is a bundle with contractible fibers, the smooth bundle

$$Y \to D|T^i \to T^i$$

is fiber diffeomorphic to $\pi^*(D|\Sigma^i)$, i.e., we have a smooth family of diffeomorphisms

$$\phi_w : Y_w \xrightarrow{\approx} Y_s \quad (s = \pi(w))$$

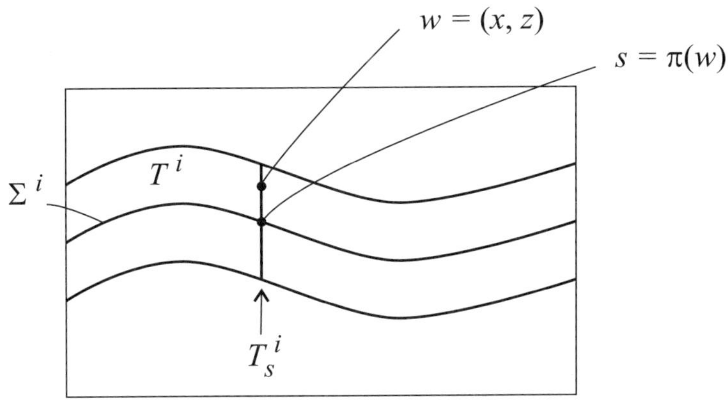

FIGURE 2. $w = (x, z) \in T^i_s = \pi^{-1}(s)$ in E (box).

for all $w \in T^i$ so that $\phi_w = \phi_s = id$ if $w = s \in \Sigma^i$. These diffeomorphisms give

$$D|T^i_s \cong \epsilon D^i \times \epsilon D^{n-i}_s \times Y_s \tag{3.13}$$

where $T^i_s = \pi^{-1}(s)$ and $Y_s = D|s$ is the fiber of D over $s \in \Sigma^i$.

Let $g : D|\Sigma \to [1, 3]$ be a fiberwise framed function, i.e., giving a family of framed functions

$$g_s : (Y_s, \partial Y_s) \to ([1, 3], 3), \quad s \in \Sigma^i$$

with critical values in $[1, 2]$. Then g extends to a fiberwise framed function on $D|T$ ($T = \coprod T^i$) by

$$g_w = g_s \circ \phi_w : Y_w \to Y_s \to I. \tag{3.14}$$

Let $\overline{g} : D \to I$ be any smooth extension of g satisfying (3.14) for all $w \in T$.

Let $h : D \to \mathbb{R}$ be given by

$$h(y) = f(w) + \delta\overline{g}(y)$$

if $y \in Y_w$ where $0 < \delta < \epsilon^2$ is fixed. If $y \in D|T^i$ then, combining (3.12) and (3.14) we have, in terms of the local coordinates (3.13),

$$h(x, z, y) = f(s) - \|x\|^2 + \|z\|^2 + \delta g_s(y).$$

The fiberwise gradient of h (over B) is equal to zero (on $D|T^i$) iff $x = 0$, $z = 0$ and $y \in \Sigma(g)$.

Adding the tangent vectors $\nabla x_1, \cdots, \nabla x_i$ in front of the framing vectors for g, we obtain a framed structure on h restricted to $D|T$. If $\delta > 0$ is chosen sufficiently small then h will have no other critical points. (The gradient of f outside of T is bounded below and the gradient of $\overline{g}$ is bounded above.) Consequently, h is fiberwise framed with the same critical set as g:

$$\Sigma(h) = \Sigma(g).$$

For each i, the fiberwise framed function g on $D|\Sigma^i$ gives an expansion functor

$$\xi_i : \text{simp}\,\Sigma^i \to \mathcal{W}h^h_\bullet(M_r(\mathbb{C}), U(r)),$$

where r is the dimension of the fibers of $\mathcal{F}$. Since $D|T$ is a convex set equivalent to the main example (3.22), the local suspension lemma (3.27) tells us that $\Sigma^i(\xi_i)$ is a local subquotient functor of the cellular chain complex functor of h. Since $\Sigma^i(\xi_i)$

is acyclic, we can use the local splitting lemma as in Proposition 3.26 (acyclic local subquotients) to simplicially homotope $C_*(h;\mathcal{F})$ into a local direct sum:

$$C_*(h;\mathcal{F}) \simeq \bigoplus_{i=0}^{n} tr_B^{\Sigma^i}(\Sigma^i(\xi_i)). \tag{3.15}$$

Equation (3.15) proves the theorem since, by additivity, suspension theorem and the "transfer commutes with torsion" formula (3.9), we have:

$$\begin{aligned}
\tau_k\left(\bigoplus tr_B^{\Sigma^i}(\Sigma^i(\xi_i))\right) &= \sum tr_B^{\Sigma^i}(\tau_k(\Sigma^i(\xi_i))) \\
&= \sum (-1)^i tr_B^{\Sigma^i}(\tau_k(\xi_i)) \\
&= \sum (-1)^i tr_B^{\Sigma^i}(\tau_k(D|\Sigma_i;\mathcal{F})_{\Sigma^i}).
\end{aligned}$$

□

CHAPTER 4

The Framing Principle

This chapter gives a precise statement of the Framing Principle.

4.1. Statement for Morse bundles

We first state the Framing Principle in the case when there is a fiberwise oriented Morse function $f : E \to \mathbb{R}$. In that case the singular set $\Sigma(f)$ is a finite covering space of B. It is a disjoint union of covering spaces

$$\Sigma(f) = \coprod \Sigma^i(f).$$

For each $i \geq 0$, we have an i–dimensional vector bundle γ^i over $\Sigma^i(f)$ given by the negative eigenspace of the second derivative of f (equal to the tangent plane to the descending manifold).

If each of these vector bundles γ^i is trivial then f admits a framed structure so it can be used to define the higher torsion of the smooth bundle E.

$$\tau_{2k}(E) := \tau_{2k}(C_*(f)) \in H^{4k}(B;\mathbb{R})$$

where $\tau_{2k}(C_*(f)) := \tau_{2k}(C(C_*(f)))$ is the torsion of the canonical cone (Lemma 2.35) of the family of chain complexes defined by f. If the bundles γ^i are nontrivial then this formula must be modified.

$$\tau_{2k}(E) = \tau_{2k}(C_*(f)) + \sum_{i\geq 0}(-1)^{i+k}\zeta(2k+1)\left(p|\Sigma^i(f)\right)_*\left(\tfrac{1}{2}ch_{2k}(\gamma^i\otimes\mathbb{C})\right) \tag{4.1}$$

where ch_{2k} is the degree $4k$ component of the *Chern character* and

$$\left(p|\Sigma^i(f)\right)_* : H^*(\Sigma^i(f);\mathbb{R}) \to H^*(B;\mathbb{R})$$

is the *push-down* operator given by summing over the sheets of the covering:

$$\left\langle \left(p|\Sigma^i(f)\right)_*(c), \sigma\right\rangle = \sum_{\widetilde{\sigma}} \langle c, \widetilde{\sigma}\rangle\,.$$

(A cocycle c is evaluated on a singular simplex σ in B by taking the sum of the values of c on the liftings $\widetilde{\sigma}$ of σ to $\Sigma^i(f)$.)

Equation (4.1) is called the *Framing Principle.* Thus equation (4.1) says that the Framing Principle holds for fiberwise oriented Morse functions.

The basic example of the use of the Framing Principle is the following (from [**Igu02**]). U. Bunke [**Bun**] had already proven the analogous statement for higher analytic torsion.

THEOREM 4.1 (torsion of sphere bundles). *Let $S^{n-1} \to E(\xi) \to B$ be the sphere bundle of an oriented n–plane bundle ξ over B. Then*

$$\tau_{2k}(E(\xi)) = (-1)^{k+n-1}\zeta(2k+1)\left(\tfrac{1}{2}ch_{2k}(\xi\otimes\mathbb{C})\right).$$

REMARK 4.2 (relation to Pontrjagin classes). To express this formula in terms of the Pontrjagin classes $p_i(\xi)$ of ξ (as in the introduction), we need to substitute

$$\tfrac{1}{2}ch_{2k}(\xi \otimes \mathbb{C}) = \frac{1}{(2k)!}N_k(p_1(\xi), p_2(\xi), \cdots)$$

where N_k is the k–th *Newton polynomial*, i.e., the integer polynomial which expresses the power sum

$$\sum x_i^k$$

in terms of the elementary symmetric functions of x_i.

Conversely, the Pontrjagin classes $p_j(\xi)$ can be expressed in terms of the higher FR–torsion classes $\tau_{2k}(E(\xi))$ by

$$\begin{aligned}\sum_{j\geq 0} p_j(\xi) &= \prod_{k>0} \exp\left(\frac{(-1)^{k-1}(2k)!}{k}\tfrac{1}{2}ch_{2k}(\xi \otimes \mathbb{C})\right) \\ &= \prod_{k>0} \exp\left(\frac{(-1)^n(2k)!}{k\zeta(2k+1)}\tau_{2k}(E(\xi))\right).\end{aligned}$$

Therefore,

$$\tau_{2k}(E(\xi)) = (-1)^n \frac{k\zeta(2k+1)}{(2k)!}p_k(\xi) + \text{higher terms}$$

where the "higher terms" add up to a polynomial in the lower Pontrjagin classes.

REMARK 4.3 (logic is reversed). This theorem is actually a lemma in the proof of the Framing Principle. The higher torsion of oriented linear sphere bundles is computed directly in [**Igu02**] and this calculation was used there to prove a restricted version of the Framing Principle. Thus, the following argument actually shows that Theorem 4.1 is equivalent to the Framing Principle in the special case of fiberwise oriented Morse functions.

PROOF. Let $D(\xi)$ be the disk bundle of ξ. Then $E(\xi) = \partial D(\xi)$. So, Corollary 3.18 gives:

$$\tau_{2k}(E(\xi)) = \tau_{2k}(D(\xi)) - \tau_{2k}(D(\xi), E(\xi)).$$

To compute $\tau_{2k}(D(\xi))$ we take the fiberwise Morse function $f : D(\xi) \to \mathbb{R}$ given by

$$f(x) = \|x\|^2.$$

This is fiberwise framed. It has a constant family of chain complexes ($0 \to \mathbb{Z} \to 0$ on every fiber). Thus

$$\tau_{2k}(D(\xi)) = 0.$$

To compute the relative torsion $\tau_{2k}(D(\xi), E(\xi))$ we take the fiberwise Morse function

$$1 - f : (D(\xi), E(\xi)) \to (I, 0).$$

The chain complex of the fiber is again constant. However, $1 - f$ is not framed since it has a critical point of index n with descending disk bundle $\gamma^n = \xi$. Thus the Framing Principle (4.1) gives:

$$\tau_{2k}(D(\xi), E(\xi)) = (-1)^{n+k}\zeta(2k+1)\left(\tfrac{1}{2}ch_{2k}(\xi \otimes \mathbb{C})\right).$$

The theorem follows. □

4.2. General statement

Suppose now that $f : (E, \partial_0 E) \to (I, 0)$ is a fiberwise oriented GMF. Then the singular set $\Sigma(f)$ is a smooth submanifold of E. It is a union of closed subsets

$$\Sigma(f) = \bigcup \overline{\Sigma}^i(f)$$

which intersect along the birth-death sets

$$\Sigma_1^i(f) = \overline{\Sigma}^i(f) \cap \overline{\Sigma}^{i+1}(f).$$

The negative eigenspace bundle γ^i is defined over all of $\overline{\Sigma}^i(f)$ and is a subbundle of γ^{i+1} over $\Sigma_1^i(f)$. The complementary line bundle is the kernel of $D^2(f)$. This has an intrinsic orientation given by the intrinsic third derivative of f. By definition of an oriented GMF, the orientation of γ^{i+1} is given by the orientation of γ^i together with the intrinsic orientation of the complementary line bundle.

DEFINITION 4.4 (stable bundle). For any oriented fiberwise GMF $f : (E, \partial_0 E) \to (I, 0)$ we define the *stable bundle* γ_f be the n–dimensional vector bundle over $\Sigma(f)$ (where $n = \dim M$) given by the following clutching construction. The restriction of γ_f to $\overline{\Sigma}^i(f)$ is equal to γ^i plus a trivial $n - i$ plane bundle spanned by vector fields $v_{i+1}, \cdots, v_n$. Along the birth-death set $\Sigma_1^i(f)$, let v_{i+1} be identified with the positive cubic direction of the kernel of $D^2(f)$ and let $v_{i+2}, \cdots, v_n$ be identified so that they extend continuously to $\overline{\Sigma}^i(f) \cup \overline{\Sigma}^{i+1}(f)$. Thus each vector field v_j is continuous over its domain.

PROPOSITION 4.5 (stable + unstable). If $f : (E, \partial_0 E, \partial_1 E) \to (I, 0, 1)$ is an oriented fiberwise GMF so that $-f$ is also oriented (i.e., the fibers M_t are oriented in a coherent way), then $\gamma_f \oplus \gamma_{-f}$ is stably isomorphic to the normal bundle of $\Sigma(f)$ in E.

REMARK 4.6 (normal=vertical tangent bundle). Since $\Sigma(f)$ is the transverse inverse image of the zero section of the vertical tangent bundle of E, the normal bundle of $\Sigma(f)$ in E is isomorphic to the restriction to $\Sigma(f)$ of the vertical tangent bundle of E.

PROOF. Along the index i Morse point set $\Sigma_0^i(f)$, the normal bundle is

$$\nu(\Sigma_0^i(f)) = \gamma_f^i \oplus \gamma_{-f}^{n-i}.$$

If we add a trivial n–plane bundle ϵ^n to $\nu(\Sigma_0^i(f))$ we get an isomorphism

$$\phi : (\gamma_f \oplus \gamma_{-f})|_{\Sigma_0^i(f)} \xrightarrow{\approx} \nu(\Sigma_0^i(f)) \oplus \epsilon^n$$

by sending the auxiliary vectors

$$v_{i+1}, v_{i+2}, \cdots, v_n, v_{n-i+1}^-, \cdots, v_n^-$$

to the basis vectors $e_{i+1}, \cdots, e_n, e_i, \cdots, e_0$, respectively. If we use the notation

$$u_j = v_{n-j+1}^-$$

then the isomorphism ϕ sends u_j to e_j and v_k to e_k.

Near the birth-death set $\Sigma_1^{i-1}(f)$ the auxiliary vectors v_i and u_i are identified with $\frac{\partial}{\partial x_i} \in \gamma_f^i$ over $\Sigma^i(f)$ and $-\frac{\partial}{\partial x_i} \in \gamma_{-f}^{n-i+1}$ over $\Sigma^{n-i+1}(-f) = \Sigma^{i-1}(f)$, respectively. (See Figure 1.)

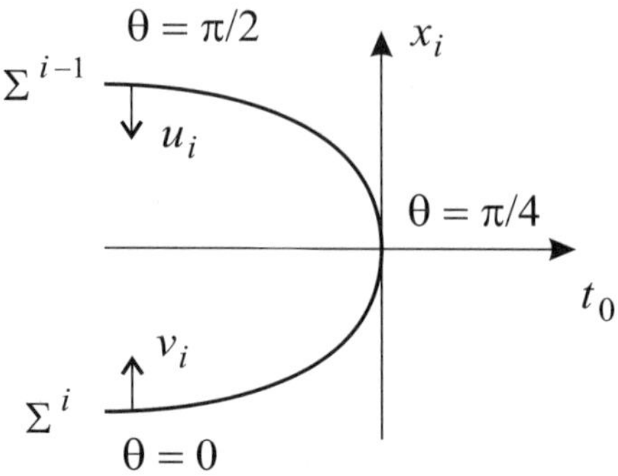

FIGURE 1. v_i rotates to u_i.

The isomorphism $\phi : \gamma_f \oplus \gamma_{-f} \to \nu(\Sigma(f)) \oplus \epsilon^n$ can thus be defined near Σ_1^{i-1} by

$$\phi(x, y, \sum a_j u_j, \sum b_k v_k) = x + y + (-a_i \sin\theta + b_i \cos\theta)\frac{\partial}{\partial x_i}$$

$$-\sin\theta\cos\theta\frac{\partial}{\partial t_0} + \sum_{j=1}^{i-1} a_j e_j + (a_i \cos\theta + b_i \sin\theta) e_i + \sum_{k=i+1}^{n} b_k e_k.$$

□

If f is a fiberwise framed function then γ_f is a trivial bundle over $\Sigma(f)$. If f is not framed then the Framing Principle says:

$$\tau_{2k}(E, \partial_0 E) = \tau_{2k}(C_*(f)) + (-1)^k \zeta(2k+1) p_* \left(\tfrac{1}{2} ch_{2k}(\gamma_f \otimes \mathbb{C})\right) \tag{4.2}$$

where the push-down operator p_* is defined below.

4.3. Push-down/transfer

In deRham cohomology, the push-down operator

$$p_*^E : H^{q+n}(E, \partial E; \mathbb{R}) \to H^q(B; \mathbb{R})$$

is given by integration along the fibers which must be oriented. (See, e.g., [**BT82**].) Thus, if σ is a smooth q–simplex in B and ω is a closed $q+n$ form on E which is zero on ∂E then the value of ω on σ is just the integral of its restriction to $p^{-1}\sigma \cong \sigma^q \times M^n$. It is well-known [**Mor01**] that this can also be given integrally as the *Gysin homomorphism*:

$$H^{q+n}(E, \partial E; \mathbb{Z}) \twoheadrightarrow E_\infty^{q,n} \hookrightarrow E_2^{q,n} \cong H^q(B; H^n(M; \mathbb{Z})) \cong H^q(B; \mathbb{Z}) \tag{4.3}$$

where we use the orientation of M to identify $H^n(M; \mathbb{Z})$ with $\mathbb{Z}$.

DEFINITION 4.7 (relative Euler class). The *relative Euler class*

$$e_{E,\partial_0} = e(E, \partial_0 E) \in H^n(E, \partial E; \mathbb{Z})$$

is defined to be the pullback of the Thom class of the verical tangent bundle $T^v E$ along any section, i.e., vertical vector field, $v : E \to T^v E$ which points inward along $\partial_0 E$ and outward along $\partial_1 E$. If $\partial_0 E$ is empty we write $e_E = e(E, \emptyset)$.

The relative Euler class depends on the choice of $\partial_0 E$. For example, $e(\partial_0 E \times I, \partial_0 E) = 0$ but $e_{\partial_0 E \times I} = \Sigma e_{\partial_0 E}$ where Σ is the isomorphism

$$\Sigma : H^{n-1}(\partial_0 E; \mathbb{Z}) \xrightarrow{\approx} H^n(\partial_0 E \times I, \partial_0 E \times \partial I; \mathbb{Z}).$$

If we compose the Gysin homomorphism with the cup product with $e_{E,\partial_0} \in H^n(E, \partial E; \mathbb{Z})$ we get the *relative transfer homomorphism*

$$tr_B^{E,\partial_0} : H^q(E;\mathbb{Z}) \xrightarrow{\cup e_{E,\partial_0}} H^{q+n}(E, \partial E; \mathbb{Z}) \xrightarrow{p_*^E} H^q(B;\mathbb{Z}) \tag{4.4}$$

which can be expressed in terms of the usual transfer by

$$tr_B^{E,\partial_0} = tr_B^E - j^* tr_B^{\partial_0 E}$$

where $j : \partial_0 E \to E$ is the inclusion map. The transfer does not require oriented fibers. It is defined when the fiber is homotopy equivalent to a finite cell complex [**BG75**].

In the Framing Principle (4.2) we use the push-down operator

$$p_*^\Sigma : H^q(\Sigma(f); \mathbb{R}) \to H^q(B; \mathbb{R}) \tag{4.5}$$

for $q = 4k$ which can be described (for all $q \geq 0$) in two ways.

The first definition is in terms of differential forms. It does not require f to be a fiberwise oriented GMF. We only need f to be a generic smooth function on E whose fiberwise gradient points inward along $\partial_0 E$ and outward along $\partial_1 E$. Since $\Sigma(f)$ is normally oriented with codimension n, it has a dual form η_Σ which is a closed n–form with support in a small tubular neighborhood T of $\Sigma(f)$ in E representing the Thom class of the normal bundle of $\Sigma(f)$ in E. We define the push-down operator (4.5) to be given on a closed q–form ω on $\Sigma(f)$ by:

$$p_*^\Sigma(\omega) = p_*^E(\pi^*(\omega) \wedge \eta_\Sigma) \tag{4.6}$$

where $\pi^*(\omega)$ is the pullback of ω along the projection $\pi : T \to \Sigma(f)$.

Although (4.6) is in terms of differential forms it actually gives an integral operation

$$p_*^\Sigma : H^q(\Sigma(f); \mathbb{Z}) \to H^q(B; \mathbb{Z}).$$

To see this note that the singular set $\Sigma(f)$ is the inverse image of the zero section of the vertical tangent bundle of E under a transverse mapping ∇f which points inward along $\partial_0 E$ and outward along $\partial_1 E$. Thus the Thom class of the normal bundle of $\Sigma(f)$ in E, which is an integral class

$$[\eta_\Sigma] \in H^n(T, \partial T; \mathbb{Z})$$

maps to the relative Euler class $e_{E,\partial_0} \in H^n(E, \partial E; \mathbb{Z})$ under the mapping

$$k^* : H^n(T, \partial T; \mathbb{Z}) \cong H^n(E, E - \mathrm{int} T; \mathbb{Z}) \to H^n(E, \partial E; \mathbb{Z})$$

given by excision and restriction. Thus (4.6) is the composition of the maps

$$H^q(\Sigma(f)) \xrightarrow{\pi^*} H^q(T) \xrightarrow{\cup[\eta_\Sigma]} H^{q+n}(T, \partial T) \xrightarrow{k^*} H^{q+n}(E, \partial E) \xrightarrow{p_*^E} H^q(B). \tag{4.7}$$

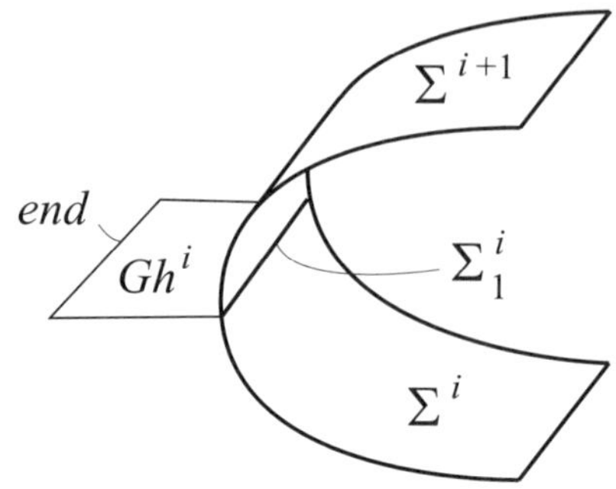

FIGURE 2. The ghost set Gh^i is attached to $\Sigma_1^i(f)$.

PROPOSITION 4.8 (push-down/transfer diagram). The following diagram commutes where i^*, j^* are restriction maps.

$$\begin{array}{ccccc}
 & & H^q(T;\mathbb{Z}) & \xrightarrow{\cup[\eta_\Sigma]} & H^{q+n}(T,\partial T;\mathbb{Z}) \\
 & \nearrow^{\pi^*} & \uparrow{\scriptstyle j^*} & & \downarrow{\scriptstyle k^*} \\
H^q(\Sigma(f);\mathbb{Z}) & \xleftarrow{i^*} & H^q(E;\mathbb{Z}) & \xrightarrow{\cup e_{E,\partial_0}} & H^{q+n}(E,\partial E;\mathbb{Z}) \\
 & \searrow_{p_*^\Sigma} & \downarrow{\scriptstyle tr_B^{E,\partial_0}} & \swarrow_{p_*^E} & \\
 & & H^q(B;\mathbb{Z}) & &
\end{array}$$

We now give another description of the integral push-down operator p_*^Σ in the case when f is a fiberwise oriented GMF. Basically, it is the alternating sum of the push-downs on $\Sigma^i(f)$, the set of index i critical points of f.

The projection map $p : E \to B$ gives a codimension one immersion on the birth-death set:

$$p : \Sigma_1^i(f) \to B.$$

This immersion is *two-sided* in the sense that its normal bundle is oriented. For t on the *inside* of $p(\Sigma_1^i(f))$ the function f_t has two cancelling critical points of index i and $i+1$. The *outward* direction is the normal direction in which these critical points vanish. However, there is still a "ghost" of a critical point which we define to be a lifting Gh^i of an exterior immersed open collar of $p(\Sigma_1^i(f))$ in B. This lifting should start at the set $\Sigma_1^i(f)$ and extend outward. (See Figure 2.)

In local coordinates we have:

$$f_t(x) = -x_1^2 - \cdots - x_i^2 + x_{i+1}^3 + t_0 x_{i+1} + x_{i+2}^2 + \cdots + x_n^2$$

where dt_0 is the outward normal direction to $p(\Sigma_1^i(f))$. For $t_0 < 0$ we have two critical points which converge to the birth-death point $x = 0$ at $t_0 = 0$. For $0 < t_0 < \epsilon$ we take the point $x = 0$ to be the ghost point. The limit points (where $t_0 = \epsilon$) form the *end* of the ghost set.

Let $\widehat{\Sigma^i}(f)$ be the union of $\overline{\Sigma}^i(f)$ with the two exterior open collars Gh^i and Gh^{i-1}:

$$\widehat{\Sigma^i}(f) = \overline{\Sigma}^i(f) \cup Gh^i \cup Gh^{i-1}.$$

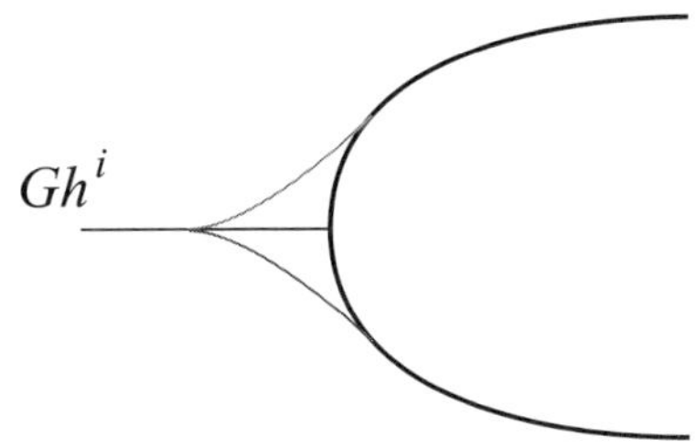

FIGURE 3. Smoothings of $\widehat{\Sigma^i}(f)$ and $\widehat{\Sigma^{i+1}}(f)$.

Then $\widehat{\Sigma^i}(f)$ is a topological manifold with the same dimension as B and

$$p : \widehat{\Sigma^i}(f) \to B$$

is locally a homeomorphism. (In Figure 2, Gh^i and Σ^i form the displayed part of $\widehat{\Sigma^i}(f)$ and $Gh^i \cup \Sigma^{i+1}$ is the displayed part of $\widehat{\Sigma^{i+1}}(f)$. The ghost set Gh^i is transverse to the singular set so $\widehat{\Sigma^i}(f), \widehat{\Sigma^{i+1}}(f)$ are not smooth manifolds.)

For any q–cocycle c^q defined in a neighborhood of $\Sigma(f)$ in E let $p_*(c)$ be the q–cocycle on B whose value on any sufficiently small (diameter $< \epsilon$) singular simplex

$$\sigma : \Delta^q \to B$$

is given by the alternating sum of the values of c on the liftings $\widetilde{\sigma}$ of σ to $\widehat{\Sigma^i}(f)$:

$$\langle p_*(c), \sigma \rangle = \sum_{i=0}^{n} (-1)^i \sum_{\widetilde{\sigma}} \langle c, \widetilde{\sigma} \rangle . \tag{4.8}$$

If $\widetilde{\sigma} : \Delta^q \to \widehat{\Sigma^i}(f)$ has image in a ghost set Gh^i or Gh^{i-1} then it is cancelled by another lifting to $\widehat{\Sigma^{i+1}}(f)$ or $\widehat{\Sigma^{i-1}}(f)$, respectively. Therefore, we may take the sum in (4.8) over only those lifting $\widetilde{\sigma}$ of σ to $\widehat{\Sigma^i}(f)$ whose image does not lie entirely in the ghost set.

PROPOSITION 4.9 (equivalence of push-downs). These two definitions (4.6), (4.8) of the push-down operator (4.5) agree.

REMARK 4.10 (orientation not needed). The *integral push-down formula* (4.8) does not require an orientation of the fibers of E.

PROOF. This is an elementary exercise. However, since it plays a crucial role in the proof of the Framing Principle, we will go through the proof quickly.

First, we smooth out the sets $\widehat{\Sigma^i}(f)$ as indicated in Figure 3 so that they become smooth manifolds locally diffeomorphic to B. We call these $\widetilde{\Sigma^i}(f)$. Since the smooth versions $\widetilde{\Sigma^i}(f)$ of the topological manifolds $\widehat{\Sigma^i}(f)$ are vertical deformations of these topological local sections with differences only in an arbitrarily small neighborhood of the birth-death set of f, we may use lifting of σ to $\widetilde{\Sigma^i}(f)$ instead of $\widehat{\Sigma^i}(f)$ in the formula (4.8) without changing the cohomology class of $p_*(c)$. As before, the terms in which $\widetilde{\sigma}$ lie entirely in the ghost set cancel.

Now suppose that the cocycle c^q in (4.8) is given by integration of the q-form $\pi^*\omega$ in (4.6). Since each $\widetilde{\Sigma^i}(f)$ is a smooth submanifold of E, it has a dual n–form η_i which is defined in the complement of the ends of the ghost sets Gh^i and Gh^{i-1}

and which has support in T (the tubular neighborhood of $\Sigma(f)$). This form has the property that

$$\sum_{\widetilde{\sigma}} \langle c, \widetilde{\sigma} \rangle = \int_{\sigma^* E} \pi^* \omega \wedge \eta_i + \langle b_i, \partial \widetilde{\sigma} \rangle \tag{4.9}$$

where b_i is a $q-1$ form which is equal to b_{i+1} on $\widetilde{\Sigma^i}(f) \cap \widetilde{\Sigma^{i+1}}(f)$.

Since the manifolds $\widetilde{\Sigma^i}(f)$ and $\widetilde{\Sigma^{i+1}}(f)$ agree near the end of Gh^i, the forms η_i and η_{i+1} can be chosen to agree near the end of Gh^i. Consequently, the alternating sum

$$\eta = \sum_{i=0}^{n} (-1)^i \eta_i$$

is a well-defined n–form on all of E and it has support in the tubular neighborhood T of $\Sigma(f)$. Combining (4.9) and (4.8) we get

$$\langle p_*(c), \sigma \rangle = \int_{\sigma^* E} \pi^* \omega \wedge \eta + \langle b, \partial \sigma \rangle \tag{4.10}$$

where the value of b^{q-1} on a small singular $q-1$ simplex τ in B is given by the alternating sum of the values of b_i on the liftings $\widetilde{\tau}$ of τ to $\widetilde{\Sigma^i}(F)$:

$$\langle b, \tau \rangle = \sum_{i=0}^{n} (-1)^i \sum_{\widetilde{\tau}} \langle b_i, \widetilde{\tau} \rangle .$$

The bundle map $\pi : T \to \Sigma(f)$ can be chosen to be vertical (parallel to the fibers of E) except near the birth-death set $\Sigma_1(f)$. Consequently, we may choose η_Σ to be equal to η outside a tubular neighborhood of $\Sigma_1(f)$. Since $\Sigma_1(f)$ has codimension $n+1$, the closed n–forms η_Σ and η must differ by an exact form $d\alpha$. So

$$\int_{\sigma^* E} \pi^* \omega \wedge \eta = \int_{\Delta^q} \sigma^* p_*^E(\pi^* \omega \wedge \eta) = \int_{\Delta^q} \sigma^* p_*^E(\pi^* \omega \wedge \eta_\Sigma) + \langle a, \partial \sigma \rangle$$

where a is given by integrating $p_*^E(\pi^* \omega \wedge \alpha)$.

Combining this with (4.10) we see that the q–cocycle $p_*(c) - \delta(a+b)$ is given by integrating the q–form $p_*^\Sigma(\omega)$ of (4.6). □

4.4. The Framing Principle

We can now state the general Framing Principle.

THEOREM 4.11 (Framing Principle). *Suppose that $p : E \to B$ is a smooth bundle with compact fiber M and $\mathcal{F}$ is an r–dimensional Hermitian coefficient system over E so that either*

a) *$\pi_1 B$ acts trivially on $H_*(M, \partial_0 M; \mathcal{F})$ or*
b) *$r = 1$ and $H_*(M, \partial_0 M; \mathcal{F})$ is $\pi_1 B$–upper triangular.*

Suppose also that $f : (E, \partial_0) \to (I, 0)$ is an oriented fiberwise GMF. Then the higher relative FR–torsion invariants $\tau_*(E, \partial_0 E; \mathcal{F})$ of $(E, \partial_0 E)$ with coefficients in $\mathcal{F}$ differs from the higher torsion given by f in the following way.

$$\tau_{2k}(E, \partial_0 E; \mathcal{F}) = \tau_{2k}(C_*(f; \mathcal{F})) + r(-1)^k \zeta(2k+1) p_*^\Sigma \left(\tfrac{1}{2} ch_{2k}(\gamma_f \otimes \mathbb{C})\right)$$

where γ_f *is the negative eigenspace bundle over* $\Sigma(f)$ *(Definition 4.4) and* p_*^Σ *is the push-down operator defined in the previous section. The correction term is zero in the other degrees:*

$$\tau_{2k+1}(E, \partial_0 E; \mathcal{F}) = \tau_{2k+1}(C_*(f; \mathcal{F})).$$

CHAPTER 5

Proof of the Framing Principle

This chapter contains the proof of both the general Framing Principle and the transfer theorem for higher torsion. These theorems are closely related and have virtually the same proof. We will first state and prove the transfer theorem. Then we modify the prove of the transfer theorem to give a proof of the Framing Principle.

5.1. Transfer theorem

Assume as before that $p : E \to B$ is a smooth bundle with compact fiber M where B is a closed manifold. Then we define the *transfer*

$$tr_B^E : H^q(E;\mathbb{Z}) \to H^q(B;\mathbb{Z})$$

from E to B as follows. If there exists a fiberwise framed function $f : E \to \mathbb{R}$ we let tr_B^E be the composition

$$H^q(E;\mathbb{Z}) \xrightarrow{i^*} H^q(\Sigma(f);\mathbb{Z}) \xrightarrow{p_*^\Sigma} H^q(B;\mathbb{Z})$$

of restriction to $\Sigma(f)$ and the integral push-down operator p_*^Σ given in (4.8). If the fibers of E are oriented, this is the same as the previously defined transfer map by the push-down/transfer diagram (Proposition 4.8).

If E is not fiberwise oriented and there is no fiberwise framed function on E then take a linear disk bundle $D^N \to D \to E$ where N is large enough to admit a fiberwise framed function $g : D \to \mathbb{R}$ and pull back the transfer from D:

$$H^q(E;\mathbb{Z}) \cong H^q(D;\mathbb{Z}) \xrightarrow{tr_B^D} H^q(B;\mathbb{Z}).$$

To see that this is independent of the choice of D, suppose that D' is another linear disk bundle over E. Then g gives a fiberwise framed function on the fiber product of D and D' over E:

$$h : D \times_E D' \to \mathbb{R}, \quad h(x, y) = g(x) + \|y\|^2.$$

But $\Sigma(h) = \Sigma(g) \times 0$ so the transfer from D to B is the pullback of the transfer from $D \times_E D'$ to B. By symmetry the same holds for D' so tr_B^D and $tr_B^{D'}$ pull back to the same mapping on $H^q(E;\mathbb{Z})$.

Suppose that D is a smooth bundle over E with fiber Y. Suppose that $\mathcal{F}$ is a Hermitian coefficient system on D so that $H_*(Y;\mathcal{F}) = 0$. Let F be the fiber of D

over B. Then F is also a bundle over M with fiber Y.

$$\begin{array}{ccccc}
Y & \longrightarrow & F & \longrightarrow & M \\
\| \downarrow & & \downarrow & & \downarrow \\
Y & \longrightarrow & D & \longrightarrow & E \\
& & \downarrow & & \downarrow \\
& & B & \xrightarrow{=} & B
\end{array}$$

Then $H_*(F;\mathcal{F}) = H_*(M; H_*(Y;\mathcal{F})) = 0$ so both terms in the following formula are define.

THEOREM 5.1 (transfer theorem). *Let $\tau_k(D;\mathcal{F})_E \in H^{2k}(E;\mathcal{F})$ denote the higher torsion of D as a bundle over E and similarly for E replaced with B. Then*

$$\tau_k(D;\mathcal{F})_B = tr_B^E(\tau_k(D;\mathcal{F})_E). \tag{5.1}$$

In the special case when D is a fiber product $D = E \times_B E'$ we get the following.

COROLLARY 5.2 (fiber product formula). Let $E' \to B$ be a smooth bundle with fiber Y. Let $\mathcal{F}$ be a Hermitian coefficient system on E' so that $H_*(Y;\mathcal{F}) = 0$ and let $\mathcal{F}'$ be the pullback of $\mathcal{F}$ to the fiber product $E \times_B E'$. Then

$$\tau_k(E \times_B E';\mathcal{F}') = \chi(M)\tau_k(E';\mathcal{F}).$$

PROOF. By the transfer theorem we have

$$\tau_k(E \times_B E';\mathcal{F}')_B = tr_B^E(\tau_k(E \times_B E';\mathcal{F}')_E) = tr_B^E p^*(\tau_k(E';\mathcal{F})).$$

Then, use the following well-known fact which follows trivially from the integral push-down formula. □

LEMMA 5.3 (transfer of pullback). *The composition*

$$H^q(B;\mathbb{Z}) \xrightarrow{p^*} H^q(E;\mathbb{Z}) \xrightarrow{tr_B^E} H^q(B;\mathbb{Z})$$

is multiplication by $\chi(M)$.

5.2. Stratified deformation lemma

The first step in the proof of the transfer theorem is to stabilize so that we can use the Framed Function Theorem. To make $\dim M > \dim B$ we replace E with a disk bundle E' with the corners rounded. Then we have an isomorphism

$$\phi^* : H^{2k}(E) \xrightarrow{\approx} H^{2k}(E')$$

with any coefficients. Let D' be the pullback of D to E'. Then

$$tr_B^{E'}\tau_k(D';\mathcal{F}')_{E'} = tr_B^{E'}\phi^*(\tau_k(D;\mathcal{F})_E) = tr_B^E(\tau_k(D;\mathcal{F})_E)$$

and

$$\tau_k(D';\mathcal{F}')_B = \tau_k(D;\mathcal{F})_B$$

by the product formula (Lemma 3.28).

To make $\dim Y > \dim B$ we take the product of Y with an even dimensional sphere. The torsion of $D \times S^{2N}$ is twice the torsion of D over both E and B so the transfer equation for $D \times S^{2N}$ is equivalent to the one for D.

Now that $\dim M > q = \dim B$, there is a fiberwise framed function

$$f : E \to \mathbb{R}.$$

The fiberwise singular set $\Sigma = \Sigma(f)$ is a smooth closed q–submanifold of E which is *stratified* by index and *codimension.* (Birth-death points have codimension 1. Morse points have codimension 0.) The restriction of $p : E \to B$ to Σ gives a smooth mapping $\Sigma \to B$ which is a local diffeomorphism on the codimension 0 set and has *fold singularities* along the codimension 1 set $\Sigma_1 = \Sigma_1(f)$. In other words, it has the local form

$$(x_1, x_2, \cdots, x_q) \mapsto (x_1^2, x_2, \cdots, x_q)$$

where Σ_1 is given by $x_1 = 0$, Σ^i is given by $x_1 \leq 0$ and Σ^{i+1} is given by $x_1 > 0$.

By a *stratified deformation* of Σ over B we mean a compact $q+1$ dimensional smooth manifold W over $B \times I$ stratified by index and codimension as described above so that $\partial_0 W = W|B \times 0$ and $\partial_1 W = W|B \times 1$ are stratified closed manifolds over B and $\partial W = \partial_0 W \coprod \partial_1 W$.

The restriction of $Y \to D \to E$ to Σ is a smooth bundle

$$Y \to D|\Sigma \to \Sigma$$

which is classified by a continuous map

$$\psi_\Sigma : \Sigma \to BDiff(Y; \mathcal{F})$$

where $Diff(Y;\mathcal{F})$ is the group of diffeomorphisms ϕ of Y together with an automorphism $\widetilde{\phi}$ of the Hermitian coefficient system $\mathcal{F}$ covering ϕ (i.e., $\widetilde{\phi} : \mathcal{F} \cong \phi^* \mathcal{F}$).

Note that ψ_Σ is the restriction to Σ of the mapping

$$\psi_E : E \to BDiff(Y; \mathcal{F})$$

which classifies the bundle $Y \to D \to E$. The higher torsion class $\tau_k(D; \mathcal{F})_E$ is the pullback along ψ_E of the *universal higher torsion class*

$$\tau_k \in H^{2k}(BDiff(Y; \mathcal{F}); \mathbb{R}).$$

By a *stratified deformation* of the *singular pair* (Σ, ψ_Σ) we mean a stratified deformation W of Σ as above together with a continuous mapping

$$\psi_W : W \to BDiff(Y; \mathcal{F})$$

extending ψ_Σ.

LEMMA 5.4 (stratified deformation lemma). *Both sides of the transfer formula (5.1) are invariants of the stratified deformation class of the singular pair* (Σ, ψ_Σ)*. Furthermore, they are both additive with respect to disjoint union of such pairs.*

REMARK 5.5 (Becker-Gottlieb transfer). This lemma can be interpreted as saying that both sides of the transfer formula depend only on the homotopy class of the mapping:

$$B \to Q(E_+) \to Q(BDiff(Y; \mathcal{F})_+)$$

where the first map $B \to Q(E_+)$ is the *Becker-Gottlieb transfer* [**BG75**].

PROOF. Case 1 (algebraic torsion). First consider the right hand side (RHS) of the transfer formula (5.1). By definition of the transfer tr_B^E, the RHS is given by the push-down of the pullback of the universal torsion class τ_k:

$$tr_B^E(\tau_k(D; \mathcal{F})_E) = p_*^E \psi_E^*(\tau_k) = p_*^\Sigma \psi_\Sigma^*(\tau_k).$$

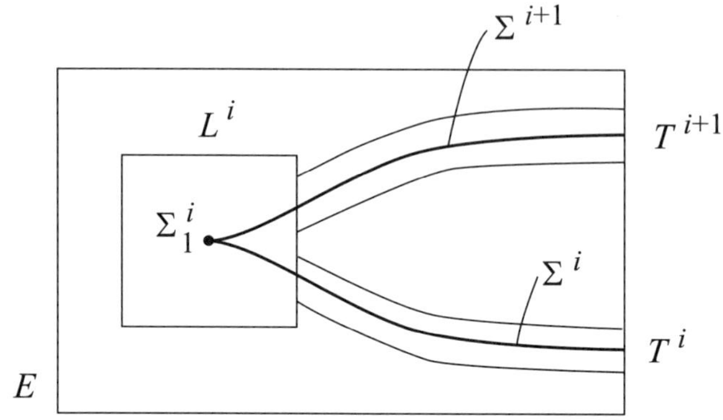

FIGURE 1. L^i (square) envelops the b-d set Σ_1^i.

We call this the *algebraic torsion* of (Σ, ψ_Σ). To easily see that this is a stratified deformation invariant, let (W, ψ_W) be any stratified deformation of (Σ, ψ_Σ). Then

$$p_*^\Sigma \psi_\Sigma^*(\tau_k) = p_*^W \psi_W^*(\tau_k)|B \times 0 = p_*^W \psi_W^*(\tau_k)|B \times 1$$

where $p_*^W : H^{2k}(W;\mathbb{R}) \to H^{2k}(B \times I;\mathbb{R})$ is given by the integral push-down formula (after attaching an ghost set along W_1). Additivity with respect to disjoint union is obvious.

Case 2 (topological torsion). Now take the left hand side (LHS) of the transfer formula. This side is defined topologically, using the Framed Function Theorem.

Case 2a (f is Morse). Suppose first that Σ_1 is empty, i.e., f has no birth-death singularities. Then the statement follows from Theorem 3.36. We review the argument briefly since we need to generalize it.

We took tubular neighborhoods $T^i \subseteq E$ of the singular sets $\Sigma^i(f)$ and constructed a fiberwise framed function $h : D \to \mathbb{R}$ so that $D|T^i$ is a convex set equivalent to the main example (3.22). By the local equivalence lemma (3.20), this identifies a subquotient functor of the cellular chain complex functor for h. Since Y is acyclic, the local splitting lemma (3.13) applies. So, $C_*(h;\mathcal{F})$, whose torsion is the LHS of (5.1), locally decomposes as a direct sum of subquotient functors so it is isomorphic to the covering space push-down of the suspensions of the cellular chain complex of the fiberwise framed function $g : D|\Sigma \to \mathbb{R}$.

$$C_*(h;\mathcal{F}) \simeq \bigoplus p_*^\Sigma(\Sigma^i(C_*(g|\Sigma^i))).$$

Thus, by the suspension theorem (3.6), we have

$$\tau_k(D;\mathcal{F})_B = \tau_k(C_*(h;\mathcal{F})) = p_*^\Sigma\left(\sum(-1)^i\psi_\Sigma^*(\tau_k)\right).$$

This is a stratified deformation invariant, proving the lemma (and the theorem) in this case.

Case 2b (f has b-d points.) In this case we use two kinds of convex sets. We envelop the b-d sets $\Sigma_1^i(f)$ in subsets L^i and the Morse points $\Sigma_0^i(f)$ in subsets T^i as in the proof of Case 2a (Theorem 3.36). (See Figure 1.)

Let L^i be a small tubular neighborhood in E of the index i birth-death set $\Sigma_1^i = \Sigma_1^i(f)$. Let $\pi : L^i \to \Sigma_1^i$ be the projection map. Since f is fiberwise framed with index i along Σ_1^i, we can choose L^i to have a twisted product structure

$$L^i = \Sigma_1^i \times J \times \epsilon D^i \times_{\Sigma_1^i} \epsilon D(\gamma') \times J$$

where ϵD^i is the i–disk of radius ϵ and $\epsilon D(\gamma')$ is the radius ϵ disk bundle of the unstable bundle of f over Σ_1^i and $J = \epsilon D^1 = [-\epsilon, \epsilon]$.

Then L^i is a bundle over the codimension 0 immersed submanifold $\Sigma_1^i \times J$ of B where the fiber over $(t, u) \in \Sigma_1^i \times J$ is

$$L^i_{t,u} = \epsilon D^i \times \epsilon D(\gamma') \times J.$$

Using the parametrized Morse lemma we may assume that $f|T$ is given by

$$f_{t,u}(x, z, w) = c_t - \|x\|^2 + \|z\|^2 + w^3 - uw.$$

Since $\dim Y > q = \dim \Sigma$, there is a fiberwise framed function $g : D|\Sigma_1 \to [0, 3]$ which we write as a family of framed functions

$$g_t : (F_t, \partial F_t) \to ([0, 3], 3)$$

on the fibers F_t of D over $t \in \Sigma_1$. We can assume that the critical values of g_t lie in the open interval $(1, 2)$ for all $t \in \Sigma_1$. We can extend g_t to $L = \coprod L^i$ by composing with the projection $\pi : L^i \to \Sigma_1^i$. Then we can extend to a fiberwise framed function over the entire singular set $\Sigma = \Sigma(f)$ and then smoothly to all $t \in E$ but g_t will only be framed in a small neighborhood of Σ.

The fiber of D over $(t, u) \in \Sigma_1^i \times J$ is

$$D_{t,u} = Y_t \times L^i_{t,u} = Y_t \times \epsilon D^i \times \epsilon D(\gamma') \times J.$$

Thus $h : D \to \mathbb{R}$ can be given by $h = \delta g + f$, i.e.,

$$h_{t,u}(y, x, z, w) = \delta g_t(y) + c_t - \|x\|^2 + \|z\|^2 + w^3 - uw.$$

If δ is sufficiently small, $h_{t,u}$ will not have singularities outside of the $D|\Sigma$.

As in Example 3.24, we need to modify the function $h_{t,u}$ in a small neighborhood of the bad set where u, x, z, w are all zero. Since we have stabilized our bundles, we can keep x, z fixed and deform the function

$$h_{t,u}(y, 0, 0, w) = \delta g_t(y) + c_t + w^3 - uw$$

so that it is fiberwise framed. Adding back the terms $-\|x\|^2 + \|z\|^2$, we get a fiberwise framed function which is locally equivalent to Example 3.24.

On the rest of the singular set $\Sigma = \Sigma(f) \subseteq E$ we construct tubular neighborhoods T^i as in the proof of Case 2a (Theorem 3.36).

Since Y is acyclic, the expansion functor given by h locally decomposes as a direct sum of the local subquotients. These in turn depend only on the stratified set Σ together with the bundle $D|\Sigma$ and the coefficient system $\mathcal{F}$. In other words, it depends only on the singular pair (Σ, ψ_Σ). A stratified deformation of this pair carries enough information to deform the convex sets and therefore determines the local subquotients. The lemma therefore holds. □

5.3. Proof of transfer theorem

The proof of the transfer theorem now proceeds in three steps

a) Both sides of the transfer formula (5.1) are zero if $\psi_\Sigma : \Sigma \to BDiff(Y; \mathcal{F})$ is trivial.
b) The transfer formula holds if the ψ_Σ is trivial on the b-d set Σ_1.

c) The singular pair (Σ, ψ_Σ) can be deformed by a stratified deformation to a pair which lies in two indices $i, i+1$ and satisfies (b).

To clarify what we mean by "trivial," we choose a base point (each one called $*$) for every component of $BDiff(Y;\mathcal{F})$. By *trivial* we mean "Each component (of Σ or Σ_1) maps to the base point of the corresponding component of $BDiff(Y;\mathcal{F})$."

The logical order of the proof of these statements is: (c),(a),(b), but we discuss them in the order listed. So, suppose for a moment that we have already proved (c). Then (a) follows from Lemma 5.4. The reason is that, with ψ_Σ trivial, both sides of the transfer formula are invariants of the stratified deformation class of Σ which is equivalent to a mapping of B into $Q(S^0)$. Since $Q(S^0)$ is rationally trivial (except for $\pi_0(Q(S^0)) = \mathbb{Z}$ which gives the Euler characteristic of Y which must be zero since it is acyclic), some positive integer multiple of $(\Sigma, *)$ must be stratified null-deformable.

To prove the statement in the last paragraph, we use (c) which says that we may assume that Σ is in the two indices $i, i+1$. We view Σ^i, Σ^{i+1} as positive and negative particles, depending on the signs $(-1)^i$, $(-1)^{i+1}$. Then, Σ gives a mapping from B into the configuration space of positive and negative particles which are allowed to cancel two at a time. This configuration space is known to be homotopy equivalent to $Q(S^0)$ [**Car81**]. Caruso also has a short proof of the unstable version of this statement.

Now we prove (b) assuming (a). The proof is in two steps. First, we show that the transfer formula holds for what we call "immersed lenses." Then, we show that the singular pair (Σ, ψ_Σ) has a stratified deformation to a linear combination of such "immersed lenses" plus a singular pair with trivial ψ_Σ.

Suppose that V is a compact connected q–manifold with boundary which is immersed in the closed q–manifold B. Let $\psi_1, \psi_2 : V \to BDiff(Y;\mathcal{F})$ be two mappings which are trivial (map to the base point $*$ of the same component) on ∂V. Then the *immersed lens*

$$L_i(V, \psi_1, \psi_2) = (L, \psi_L)$$

is given by letting L be the double of V (two copies glued along ∂V) with indices $i, i+1$ where ψ_L is equal to ψ_1 on L^i and ψ_2 on L^{i+1}.

LEMMA 5.6 (immersed lenses). *The transfer formula holds for immersed lenses.*

PROOF. First, consider the case when V is embedded in B. Then we can, by a stratified deformation, eliminate all the b-d points of L. (Just double the set $V \times I \cup B \times [\frac{1}{2}, 1]$ with corners rounded.) By Theorem 3.36, the transfer formula holds.

For the general case, we use the push-down homomorphism

$$H^{2k}(V, \partial V; \mathbb{R}) \to H^{2k}(B; \mathbb{R})$$

given by the same formula as the transfer for covering maps (3.8). The algebraic and topological torsion of the singular pair (L, ψ_L) are push-downs of the corresponding torsion classes in the relative cohomology of $(V, \partial V)$ given by the identity embedding $V \to V$. Therefore, it suffices to show that the transfer formula holds for an embedded lens in the relative case.

Now, we view the singular pair (L, ψ_L) over $(V, \partial V)$ as a mapping

$$V/\partial V \to Q(BDiff(Y;\mathcal{F})_+).$$

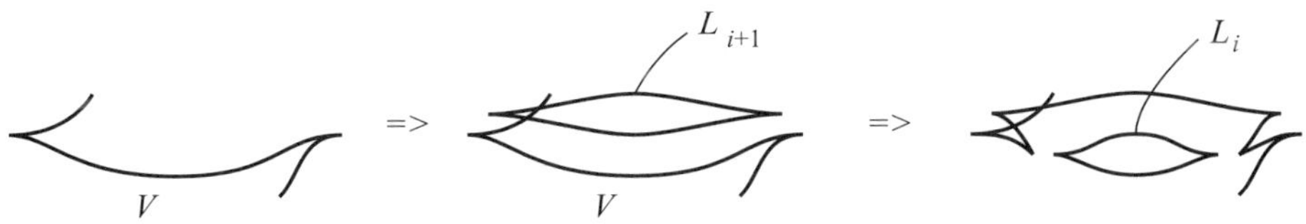

FIGURE 2. Isolate ψ_V in a lens.

By changing this mapping on the base point we can eliminate the birth-death points. This reduces us to Case 2a where the transfer formula already holds. □

To complete the proof of (b), we need the following lemma. The wording of the lemma assumes that we are working in the abelian group generated by singular pair with addition given by disjoint union. The additivity of both sides of the transfer formula under disjoint union allows us to do this.

LEMMA 5.7 (decomposition into lenses). *Any singular pair (Σ, ψ_Σ) over B so that ψ_Σ is constant on the b-d set Σ_1 is equivalent to a linear combination of immersed lenses and a trivial pair (where ψ_Σ is trivial).*

PROOF. The singular pair (Σ, ψ_Σ) will have a finite number of Morse components on which ψ_Σ is nontrivial. If this number is zero then the pair is trivial. So, it suffices to reduce this number by one by adding and deleting immersed lenses.

Let V be a component of Σ^i on which ψ_Σ is nontrivial. So

$$\psi_V = \psi_\Sigma | V : V/\partial V \to BDiff(Y; \mathcal{F})$$

is essential. Then we can add a trivial immersed lens $L_{i+1}(V, *, *)$ (being stratified null deformable) and cancel along a parallel copy of ∂V in the interior of V as indicated in Figure 2. This isolates ψ_V in an immersed lens isomorphic to $L_i(V, \psi_V, *)$. Remove this lens and replace it with a trivial lens $L_i(V, *, *)$. Then go backwards in the deformation of Figure 2.

The end result is that Σ is unchanged and ψ_Σ has been trivialized on V. □

The last step (c) comes logically before (a) and (b). The proof of (c) will be in two steps. First, we deform any singular pair (Σ, ψ_Σ) into two indices. Next, we deform the singular pair so that each component of Σ_1 lies in a contractible subset of Σ. It will then follow that the map ψ_Σ can be changed by a homotopy so that it is trivial on Σ_1.

To deform the singular pair (Σ, ψ_Σ) into two indices we use "twisted lenses" which are defined as follows. Let V be a compact q–manifold with boundary which is immersed in B and let $\psi_V : V \to BDiff(Y; \mathcal{F})$ be any continuous mapping. Then the *twisted lens* is given by

$$L_i(V, \psi_V, \psi_V) = (L, \psi_L)$$

where $L = L^i \cup L^{i+1}$ is the double of V and $\psi_L : L \to BDiff(Y; \mathcal{F})$ is given by ψ_V on both L^i and L^{i+1}. In other words it is the composition of the folding map $L \to V$ with ψ_V.

It is obvious that every twisted lens is stratified null deformable. (A null deformation is given by doubling $V \times [0, \frac{1}{2}] \looparrowright B \times I$.) So, we can add and delete them at will. In particular, we can add two twisted lenses $L_i(V, \psi_V, \psi_V)$ and

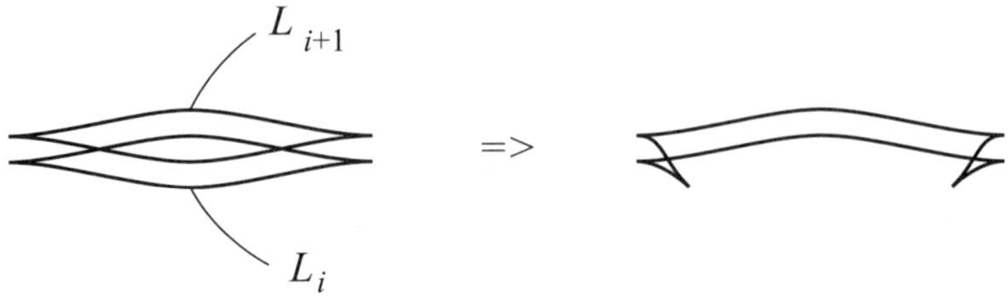

FIGURE 3. Cancel the bottoms of two twisted lenses.

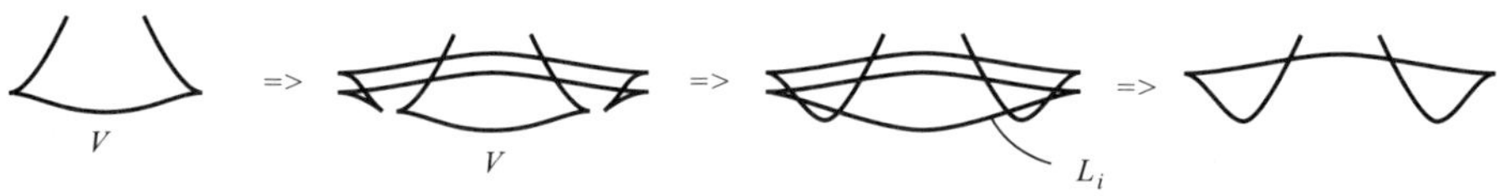

FIGURE 4. Place V in twisted lens L_i and eliminate.

$L_{i+1}(V, \psi_V, \psi_V)$. The bottom portion of each of these twisted lenses can be cancelled by a stratified deformation to form a singular pair as shown in Figure 3.

Now let i be minimal so that Σ^i is nonempty. Let V be a component of Σ^i. Let $\psi_V = \psi_\Sigma|V$. Use this to construct the singular pair in Figure 3. Then, along ∂V, the mappings to $BDiff(Y; \mathcal{F})$ agree so we can perform the stratified deformation shown in Figure 4. This places V inside a new twisted lens $L_i(V, \psi_V, \psi_V)$ which we can eliminate. The result is that the number of components of Σ^i decreases by one. (And a new component of Σ^{i+1} is introduced.) By induction, we can deform the singular pair into two indices.

Assume now that $\Sigma = \Sigma^i \cup \Sigma^{i+1}$. The final step in the proof of the transfer theorem is to show that ψ_Σ can be trivialized on Σ_1. In fact, we will arrange for every component of Σ^i to be contained in a contractible subset of Σ. Since $\Sigma_1 = \partial\Sigma^i$, this will prove the theorem.

Choose a triangulation of Σ^i so that each simplex maps monomorphically into B. Then, we will "cut apart" the set Σ^i by deleting a tubular neighborhood of each simplex starting at the lowest dimension. If we let S denote what is left of the original set Σ^i, then, at the end, the set S will be a disjoint union of q–disk $(q = \dim B)$.

At each step of the deformation, a new component of Σ^i will be introduced which will be contained in a q–disk disjoint from the set S. Every step of the deformation will alter the set Σ only in an arbitrarily small neighborhood of the set S. Consequently, the q–disks containing the new components of Σ^i will not be affected.

Now we describe the deformation. The deformation is almost the same as the one given in [**Igu87**] and illustrated in Figures E,F,G,H,I,J in [**Igu87**]. (But here we have $i+1$ instead of $i-1$.) Suppose first that v is vertex in the interior of $S \subseteq \Sigma^i$. Then we introduce a trivial lens L over the image of v in B mapping to the point $\psi_\Sigma(v)$. Then L^i is a new component of Σ^i which is contractible. We will not touch this set again. Instead we cancel part of L^{i+1} with a q–disk neighborhood of v in S.

FIGURE 5. New component of Σ^i is shaded.

Suppose by induction that all internal simplices of S of dimension $< m$ have been removed. Let D^m be an m–disk embedded in S with boundary $S^{m-1} \subseteq \partial S$. Suppose for a moment that the sphere S^{m-1} also bounds an m–disk Δ in Σ^{i+1} which lies over the image of D^m in B. Suppose also that $D^m \cup \Delta$ forms a twisted lens, i.e., the restrictions of ψ_Σ to Δ and D^m agree if we compose with a diffeomorphism $\Delta \cong D^m$. Then we can cancel Δ with D^m eliminating this m–disk from S.

In general, we do not have such a disk Δ. We need to construct it. We take a sphere S'^{m-1} in the interior of Σ^{i+1} which is parallel to S^{m-1} and lies over the image of D^m in B. Over that sphere, we create a "tube" T, a product of an $m-1$ sphere with a trivial lens of dimension $q-m+1$. Then $T^i \cong T^{i+1} \cong D^{q-m+1} \times S^{m-1}$. The mapping to $BDiff(Y;\mathcal{F})$ should agree with $\psi_\Sigma|S'^{m-1}$. Then we can cancel a tubular neighborhood of S'^{m-1} with a tubular neighborhood of the core $* \times S^{m-1}$ of T^i. Then the new component of Σ^i will be equivalent to

$$S^{m-1} \times (D^{q-m+1} - \tfrac{1}{2} D^{q-m+1}).$$

This is shown as two shaded annuli in Figure 5.

Along the inner edge of Σ_1 given by $S^{m-1} \times \frac{1}{2} e_1$ where e_1 is the unit vector in the first coordinate axis, we do surgery on this $m-1$ sphere producing an $m-1$ disk of points in both Σ^i and Σ^{i+1}. This is shown as a horizontal shaded bar in Figure 5. We extend the mapping ψ_Σ along this disk so that it matches $\psi_\Sigma|D^m$. This gives a parallel disk Δ as described in the previous paragraph. (Δ is shown as a line bisecting the shaded region in Figure 5. However, it lies in Σ^{i+1}, not in Σ^i.)

By construction, we can cancel Δ with D^m. The deformation alters the original set Σ only in a small neighborhood of D^m since half of it occurs in the set Δ which is newly added.

The new component of Σ^i lies in a set diffeomorphic to $D^{q-m+1} \times S^{m-1}$ with an m–handle added, making it contractible. (In Figure 5, this contractible set is given by adding to the shaded region the two disks of Σ^{i+1} components with boundary the two outer circles. Spin this about the central vertical axis to get the case $m = 2$.) Thus, the deformation works in the required way, proving (c) and thus the transfer theorem.

5.4. Proof of Framing Principle

Now we are ready to prove the Framing Principle. Suppose that $p : E \to B$ is a smooth bundle with $\partial E = \partial_0 E \coprod \partial_1 E$ and $\mathcal{F}$ is a r-dimensional Hermitian coefficient system over E so that the relative torsion $\tau_k(E, \partial_0 E; \mathcal{F})$ is defined, i.e., either

(1) $\pi_1 B$ acts trivially on $H_*(M; \mathcal{F})$ or
(2) $r = 1$ and $H_*(M; \mathcal{F})$ is $\pi_1 B$–upper triangular.

Let

$$f : (E, \partial_0 E) \to (I, 0)$$

be a fiberwise oriented GMF and let γ_f be the negative eigenspace bundle of f. Then γ_f is an oriented vector bundle over $\Sigma(f)$.

If γ_f is a stably trivial vector bundle then the cellular chain complex functor $C_*(f; \mathcal{F})$ gives the higher FR–torsion of $(E, \partial_0 E)$. When γ_f is nontrivial the Framing Principle (4.11) tells us the difference between $C_*(f; \mathcal{F})$ and the true higher torsion of $(E, \partial_0 E)$.

The first step in proving the Framing Principle is to stabilize the bundle γ_f and the complementary bundle γ_{-f}.

LEMMA 5.8 (reduction to stable case). *It suffices to prove the Framing Principle in the case when the fibers M_t have dimension much larger than* $\dim B$ *and where the singular set $\Sigma(f)$ is a disjoint union of* framed *and* unframed *components where the framed components admit a framed structure and the unframed components have critical points of very large index and coindex.*

PROOF. To do this without introducing technical complications such as corners, we take the product with $Y = S^{2N} \times S^{2N}$ for N large. Then the product formula (3.28) tells us that

$$\tau_k(E \times Y, \partial_0 \times Y; \mathcal{F}') = \chi(Y)\tau_k(E, \partial_0 E; \mathcal{F}) = 4\tau_k(E, \partial_0 E; \mathcal{F})$$

where $\mathcal{F}'$ is the pull back of $\mathcal{F}$ to $E \times Y$. Thus it suffices to compute the higher torsion of $(E \times Y, \partial_0 \times Y)$.

Imitating the proof of the product formula, we choose a fixed Morse function

$$g : Y \to [0, 9]$$

having exactly four critical points y_0, y_1, y_2, y_3 with critical values $g(y_i) = 3i$. Thus y_0 is a minimum, y_3 a maximum and y_1, y_2 are critical points of index $2N$.

Let

$$h : E \times Y \to [0, 10]$$

be given by $h(x, y) = f(x) + g(y)$. Then the critical points of h will lie over the points y_i since

$$\Sigma(h) = \Sigma(f) \times \Sigma(g) \subseteq \coprod E \times y_i.$$

The critical values of h_t lie in the intervals $(3i, 3i + 1)$ so $3i + 2 = 2, 5, 8$ are regular values of h_t for all $t \in B$. The sets $S(i) = h^{-1}[3i - 1, 3i + 2]$ for $i = 0, 1, 2, 3$ are convex. As in the proof of the product formula, the corresponding subquotient functors are

$$\xi_i / \xi_{i-1} \cong \Sigma^{\mathrm{ind}(y_i)} C_*(f; \mathcal{F}).$$

Since the indices of y_i are all even, the torsion of each of these subquotients is equal to the torsion of f.

By the C^1–local framed function theorem (Remark 2.5), we can approximate $h|S(i)$ by a fiberwise framed function for each i. Let $\widetilde{h}$ be the fiberwise oriented

GMF which is equal to the original function h on $S(1) \cup S(2)$ but equal to the fiber-wise framed C^1–approximation on $S(0)$ and $S(3)$. Then $\widetilde{h}$ satisfies the condition of the lemma so we may assume that the Framing Principle holds for this function.

The unframed components of the singular set $\Sigma(\widetilde{h})$ lie in $S(1)$ and $S(2)$. Since the indices of y_1, y_2 are even, the correction terms in the Framing Principle are the same for both $S(1)$ and $S(2)$ as they are for f. Call this term C_{FP}. Then the Framing Principle for $\widetilde{h}$ gives

$$4\tau_k(E, \partial_0 E; \mathcal{F}) = \tau_k(\widetilde{h}; \mathcal{F}') + 2C_{FP}. \tag{5.2}$$

By the Splitting Lemma (3.15) we have

$$\tau_k(\widetilde{h}; \mathcal{F}') = \sum_{i=0}^{3} \tau_k(\widetilde{h}|S(i); \mathcal{F}').$$

We claim that the first and last summands give twice the torsion of $(E, \partial_0 E)$. To see this, we can repeat the process above with $Y = S^{4N}$ and $g : Y \to [0, 9]$ having only two critical points with critical values $0, 9$. Then $S(0) = h^{-1}[0, 2]$ and $S(3) = h^{-1}[8, 10]$ remain unchanged. So,

$$\begin{aligned}\tau_k(\widetilde{h}|S(0); \mathcal{F}') + \tau_k(\widetilde{h}|S(3); \mathcal{F}') &= \tau_k(E \times S^{4N}, \partial_0 E \times S^{4N}; \mathcal{F}') \\ &= \chi(S^{4N})\tau_k(E, \partial_0 E; \mathcal{F}) = 2\tau_k(E, \partial_0 E; \mathcal{F}).\end{aligned}$$

Therefore, subtracting this from both sides of (5.2) we get

$$2\tau_k(E, \partial_0 E; \mathcal{F}) = \tau_k(\widetilde{h}|S(1); \mathcal{F}') + \tau_k(\widetilde{h}|S(2); \mathcal{F}') + 2C_{FP}.$$

However, on $S(1)$ and $S(2)$, $\widetilde{h}$ is equal to h which is an even suspension of the original function f. Consequently, it has the same torsion and we get

$$2\tau_k(E, \partial_0 E; \mathcal{F}) = 2\tau_k(f; \mathcal{F}) + 2C_{FP}.$$

Divide by 2 to get the Framing Principle. □

We now assume the conditions of the reduction lemma (5.8). The next step is to prove the stratified deformation lemma.

LEMMA 5.9 (stratified deformation lemma for FP). *If we rewrite the Framing Principle as*

$$\tau_k(E, \partial_0 E; \mathcal{F}) - \tau_k(f; \mathcal{F}) = C_{FP} \tag{5.3}$$

then both sides of the equation are additive stratified deformation invariants of the singular pair $(\Sigma(f), \psi_f)$ *where*

$$\psi_f = (\psi_f^-, \psi_f^+) : \Sigma(f) \to BSO \times BO$$

is the classifying map of the stable negative and positive eigenspace bundles (γ_f, γ_{-f}) *of* f. *(By* additive *we mean additive with respect to disjoint union.)*

REMARK 5.10 (comparison class). The left hand side of (5.3) is always well defined even though the individual terms may not be defined. This difference class is called the *comparison class.* More details can be found in [**Igu02**].

Suppose for a moment that the lemma holds. Then, as in the proof of the transfer theorem (5.1), we need to verify the following. (Step (a) is unnecessary.)

b) The comparison formula holds if ψ_f is trivial on the b-d set Σ_1.

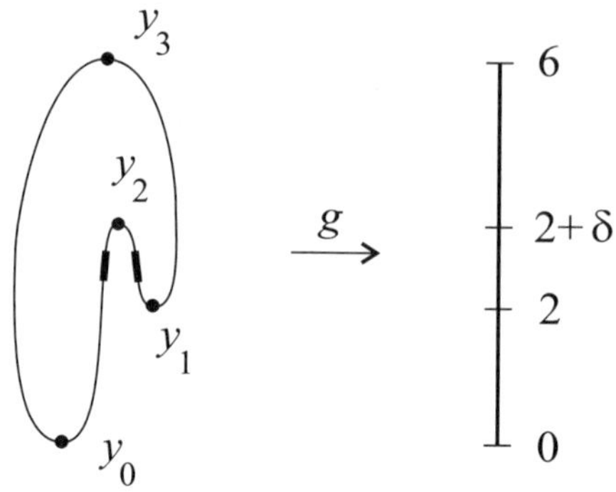

FIGURE 6. Two components of $g^{-1}[2+\frac{\delta}{3}, 2+\frac{2\delta}{3}] \subseteq S^1$ are darkened.

c) The singular pair (Σ, ψ_f) can be deformed by a stratified deformation to a pair which satisfies (b).

However, both of these conditions have already been verified. The proof of the transfer theorem shows that (c) holds for any additive stratified deformation invariant. The first statement (b) is the version of the Framing Principle proved in [**Igu02**] where we assumed that f was framed along $\Sigma_1(f)$. (This is equivalent to ψ_f^- being trivial on Σ_1 under the assumptions of the reduction lemma 5.8.) Proposition 3.4 was used in that proof.

Therefore, the general Framing Principle follows from the stratified deformation lemma above.

PROOF OF LEMMA 5.9. Since the correction term C_{FP} is a characteristic class of γ_f on Σ, it is an invariant of the stratified deformation class of (Σ, ψ_f). So, we consider only the left hand side of the comparison formula (5.3).

The argument will be the same as the proof of the product formula (3.28) and the reduction lemma. Assume that we are given a fiberwise oriented GMF

$$f : (E, \partial_0 E, \partial_1 E) \to (I, 0, 1)$$

satisfying the conditions of the reduction lemma.

Take the product of E with a circle $Y = S^1$. Let

$$g : S^1 \to [0, 6]$$

be a Morse function with four critical points y_0, y_1, y_2, y_3 having critical values $0, 2, 2+\delta, 6$, resp., where δ will be chosen later. Then these points will be arranged in the order y_0, y_2, y_1, y_3 around the circle as shown in Figure 6. Let

$$h : E \times S^1 \to [0, 7]$$

be given by $h(x, y) = f(x) + g(y)$. Then h will be a fiberwise oriented GMF. (Take any orientation of y_2, y_3.)

The vertical gradient of h with respect to the product metric is given by $\nabla h_t = (\nabla f_t, \nabla g_t)$. Consequently,

$$\Sigma(h) = \Sigma(f) \times \{y_0, y_1, y_2, y_3\}.$$

Since f is a fiberwise oriented GMF, so is h. Also, $\frac{3}{2}$ and $\frac{11}{2}$ are regular values of h_t for all $t \in B$ so $h^{-1}[\frac{3}{2}, \frac{11}{2}]$ is a convex set.

At the good points, where f is framed, so is h. Of the points where h is not framed we will modify the ones in $E \times y_1$. Then we claim that the resulting family of fiberwise oriented GMF's, $\widetilde{h}$, has the property that $\widetilde{h}^{-1}[\frac{3}{2}, \frac{11}{2}]$ is convex with acyclic subquotient functor whose torsion is the left hand side of (5.3). We prove this by taking $\delta > 1$.

We also claim that the higher torsion of $\widetilde{h}$ is a stratified deformation invariant of the singular pair $(\Sigma(f), \psi_f)$. We prove this by taking δ very small. Finally, we show that the subquotient functor of $\widetilde{h}^{-1}[\frac{3}{2}, \frac{11}{2}]$ is independent of the choice of δ. This will prove the lemma.

To prove the first assertion, take $\delta = 2$. Then $\frac{7}{2}$ will be a regular value of $\widetilde{h}_t$ for each $t \in B$. The convex set $\widetilde{h}_t^{-1}[\frac{7}{2}, \frac{11}{2}]$ will have subquotient functor equal to the suspension of $C_*(f)$ since y_2 has index 1. Also the convex set $\widetilde{h}_t^{-1}[\frac{3}{2}, \frac{7}{2}]$ will have the same torsion as the original bundle $(E, \partial_0 E)$ since it is a positive suspension of $(E, \partial_0 E)$ with a fiberwise framed function by construction of $\widetilde{h}$. Since we are assuming that this torsion is defined, we know, by the Splitting Lemma, that

$$\begin{aligned}\tau_k(\widetilde{h}|\widetilde{h}^{-1}[\tfrac{3}{2}, \tfrac{11}{2}]; \mathcal{F}) &= \tau_k(\widetilde{h}|\widetilde{h}^{-1}[\tfrac{3}{2}, \tfrac{7}{2}]; \mathcal{F}) + \tau_k(\widetilde{h}|\widetilde{h}^{-1}[\tfrac{7}{2}, \tfrac{11}{2}]; \mathcal{F}) \\ &= \tau_k(E, \partial_0 E; \mathcal{F}) - \tau_k(f; \mathcal{F})\end{aligned}$$

which is the left hand side of the comparison formula.

Next, we switch the roles of f and g. Choose tubular neighborhoods L^i for the b-d sets $\Sigma_1^i(f)$ and other neighborhoods T^i for the remaining points of $\Sigma^i(f)$ as in Figure 1 in the proof of the first stratified deformation lemma (5.4). Take $\delta < \epsilon^3/3$ where ϵ is the size of L^i. Then $L^i \times g^{-1}[2-\delta, 2+2\delta]$ is a special case of the second example of a convex set (3.24). This includes the deformation of h to $\widetilde{h}$. We also take tubular neighborhoods T^i of what is left of $\Sigma^i(f)$. Then $T^i(f) \times g^{-1}[2-\delta, 2+2\delta]$ is a special case of the main example (3.22) of a convex set for h. The same set is convex for $\widetilde{h}$, giving is a special case of Example 3.23.

Let Z be the component of $g^{-1}[2-\delta, 2+2\delta]$ containing y_1 and y_2. Then Z is contractible, so the subquotient functors given by $L^i \times Z$ and $T^i \times Z$ are acyclic and thus are locally a direct summand of the expansion functor for $\widetilde{h}$. However, $\widetilde{h}$ was a fiberwise framed function given by a small deformation of h in a neighborhood of $\Sigma(f) \times y_1$. Consequently, the subquotient functor given by the convex sets $L^i \times Z$ and $T^i \times Z$ are stratified deformation invariants. (Since the bundles are stable, they completely determine the fiber diffeomorphism type of the convex sets L^i and T^i.)

Finally, we will prove that the cellular chain complex of $\widetilde{h}$ is independent of the choice of δ. In fact we will show that the filtration of the bundle $E \times S^1$ given by $\widetilde{h}$ can be made to be independent of the choice of δ.

First we take δ small as given in the second case above. Use the standard metric on S^1 for the second case. Next we increase δ to 2 and change g so that the derivative of g is nonzero only in the set D consisting of the two darkened region in Figure 6. Thus D is the union of two of the components of $g^{-1}[2+\delta/3, 2+2\delta/3]$. Call the new function $\widetilde{g}$. Since the derivative of $\widetilde{g}$ is nowhere zero on D, we can alter the metric on S^1 so that the gradient of $\widetilde{g}$ with respect to the new metric is equal to the gradient of g with respect to the standard metric on S^1. This has no affect on our assumption that the metric be standard at all critical points since these are regular points of g and thus $E \times D$ will have no critical points of $\widetilde{h}$.

We take the product metric on $E \times S^1$ for the function h and later deform it so that $\widetilde{h}$ will have critical points with standard metric. Then we deform the metric on $E \times D$ as indicated above. This creates two functions, call them $\widetilde{h}$ and $\widetilde{h}'$ which have identical critical set and identical gradients with respect to their corresponding metrics. The minimal partial ordering of the critical set and the twisting cochain which give the expansion functors of $\widetilde{h}$ and $\widetilde{h}'$ and the subquotient functor supported on the union of the sets L^i and T^i depend only on the vertical gradient vector fields $\nabla\widetilde{h}$ and $\nabla\widetilde{h}'$. Consequently, the subquotient functors are the same. This proves the claim, the lemma and the main theorem of this paper. $\square$

CHAPTER 6

Applications of the Framing Principle

With the new Framing Principle we can compute the higher FR–torsion when the fiber is closed and even dimensional (and the higher torsion is defined). This extends the earlier calculation of the higher torsion for the Torelli group. We also give a formula for the higher torsion of a bundle with even dimensional fibers with boundary.

6.1. Torelli group

One of the main motivations for the Framing Principle was to study the higher torsion of the *Torelli group* T_g. This is the group of isotopy classes of self-diffeomorphisms ϕ of a closed oriented surface Σ_g of genus g so that ϕ induces the identity on the integral homology of Σ_g.

It was John Klein [**Kle93**] who first realized that the higher torsion invariants $\tau_{2k}(T_g)$ for the Torelli group could be defined and he conjectured that they were proportional to the Miller-Morita-Mumford class κ_{2k}.

Recall that the *Miller-Morita-Mumford classes* ([**Mum83**],[**Mor87**],[**Mil86**]) for closed oriented surface bundle $E \to B$ are the integral cohomology classes in B given by pushing down the powers of the Euler class of the vertical tangent bundle of E.

$$\kappa_k(E) := p_*^E(e_E^{k+1}) \in H^{2k}(B;\mathbb{Z}). \tag{6.1}$$

For surfaces with boundary we use the Euler class $e_E \in H^2(E,\partial E)$ to get the *punctured Miller-Morita-Mumford classes*

$$\kappa_k^\partial(E) = p_*^E(e_E^{k+1}) = p_*^E(c_1^k \cup e_E) = tr_B^E(c_1^k) \tag{6.2}$$

where $c_1 \in H^2(E)$ is the restriction of e_E to E.

Klein's conjecture was first proved by R. Hain, R. Penner and the author (unpublished) using Bökstedt's theorem [**Bök84**]. In [**Igu02**], *fat graphs* (finite graphs with cyclic orderings of the half-edges incident to each vertex) and the Framing Principle were used to find the proportionality constant. In order to use fat graphs, we needed to restrict to surfaces with at least one marked point.

THEOREM 6.1 (torsion of T_g^s). *Let T_g^s be the Torelli group of genus g surfaces with $s \geq 1$ marked points. Then*

$$\tau_{2k}(T_g^s) = \frac{1}{2}(-1)^k\zeta(2k+1)\frac{\kappa_{2k}}{(2k)!}.$$

The version of the Framing Principle proved in [**Igu02**] assumes that the fiberwise GMF $f : E \to \mathbb{R}$ is framed along the birth-death set. So, in the proof of 6.1, we needed to choose f to have maxima at the marked points and minima and saddle points on the fat graph which forms the canonical core of the complement of the set of marked points.

With the new Framing Principle we can prove this more directly without using fat graphs.

THEOREM 6.2 (torsion of oriented surface bundles). *Let $\Sigma \to E \to B$ be a closed oriented surface bundle so that the action of $\pi_1 B$ on $H_*(\Sigma, \mathbb{Q})$ is upper triangular. Then*

$$\tau_{2k}(E) = \frac{1}{2}(-1)^k \zeta(2k+1)\frac{\kappa_{2k}(E)}{(2k)!}.$$

PROOF. This is a special case of Theorem 6.6 below. □

6.2. Even dimensional fibers

The calculation of the higher torsion for surface bundles extends to all closed oriented even dimensional fibers. However, in order to state the result we need a generalization of the Miller-Morita-Mumford classes.

DEFINITION 6.3 (the integral class $T_{2k}(E)$). For any smooth bundle $E \to B$ with compact manifold fibers, let

$$T_{2k}(E) = tr_B^E\left(\frac{(2k)!}{2} ch_{2k}(T^vE \otimes \mathbb{C})\right) \in H^{4k}(B;\mathbb{Z})$$

This is an integral class by Remark 4.2.

If T^vE already has a complex structure (as in the case of oriented surface bundles), we have

$$\frac{(2k)!}{2} ch_{2k}(T^vE \otimes \mathbb{C}) = (2k)! ch_{2k}(T^vE) = \sum c_1(\lambda_i)^{2k}. \tag{6.3}$$

PROPOSITION 6.4 (T_{2k} for surface bundles). Let $\Sigma \to E \to B$ be an oriented surface bundle. Then

a) $T_{2k}(E) = \kappa_{2k}(E)$ if Σ is closed.
b) $T_{2k}(E) = \kappa_{2k}^{\partial}(E)$ if Σ has a boundary.
c) $T_{2k+1}(E) = 0$ in both cases.

PROPOSITION 6.5 (T_{2k} for M^{2n+1} oriented). If the fiber M is an oriented odd dimensional manifold then

$$2T_{2k}(E) = T_{2k}(\partial E).$$

PROOF. First suppose that M is a closed manifold of dimension $2n+1$. Let $f : E \to \mathbb{R}$ be any generic smooth function. Then the fiberwise singular set $\Sigma(f)$ is a normally oriented codimension $2n+1$ submanifold of E dual to the Euler class of T^vE. But, $\Sigma(-f) = \Sigma(f)$ with the normal orientation reversed. So

$$e_E = -e_E = 0.$$

Now suppose that M has a boundary. Then $\nabla(-f)$ points the wrong way along ∂E so we have to "fold up" a collar neighborhood of ∂E by adding the function $g(x) + t^2$ where $g : \partial E \to \mathbb{R}$ is a smooth function and t is negative the distance to ∂E. (See the proof of Corollary 3.18 for more details.) This adds the fiberwise singular set $\Sigma(g)$ to $\Sigma(-f)$. So

$$e_E = -e_E + e_{\partial E}.$$

The sign in front of $e_{\partial E}$ is the sign of the second derivative of t^2. □

Let $E^* = E \times D^N$ with the corners rounded. Then by the definition of transfer we have

$$T_{2k}(E^*) = T_{2k}(E). \tag{6.4}$$

THEOREM 6.6 (torsion for M^{2n} closed, oriented). *Suppose that M^{2n} is a closed oriented even dimensional manifold and $M \to E \to B$ is a smooth bundle where $\pi_1 B$ acts upper triangularly on the rational homology of M. (In particular, it preserves the orientation of M.) Then the higher FR–torsion invariants $\tau_{2k}(E) \in H^{4k}(B;\mathbb{R})$ are defined and given by*

$$\tau_{2k}(E) = \frac{1}{2}(-1)^k \zeta(2k+1)\frac{1}{(2k)!}T_{2k}(E).$$

PROOF. Suppose first that there is a fiberwise oriented GMF $f : E \to \mathbb{R}$. Then $-f$ is also a fiberwise oriented GMF on E so the Framing Principle gives us:

$$\tau_{2k}(E) = \tau_{2k}(C_*(f)) + (-1)^k\zeta(2k+1)p_*\left(\tfrac{1}{2}ch_{2k}(\gamma_f \otimes \mathbb{C})\right)$$

$$\tau_{2k}(E) = \tau_{2k}(C_*(-f)) + (-1)^k\zeta(2k+1)p_*\left(\tfrac{1}{2}ch_{2k}(\gamma_{-f} \otimes \mathbb{C})\right)$$

By the involution property, $\tau_{2k}(C_*(-f)) = -\tau_{2k}(C_*(f))$. Consequently, the sum of the above equations is:

$$2\tau_{2k}(E) = (-1)^k\zeta(2k+1)p_*\left(\tfrac{1}{2}ch_{2k}((\gamma_f \oplus \gamma_{-f}) \otimes \mathbb{C})\right).$$

The theorem then follows from Proposition 4.5 which implies that

$$\gamma_f \oplus \gamma_{-f} \cong T^v E|\Sigma(f) \oplus \epsilon^{2n}$$

and the push-down/transfer diagram (Proposition 4.8) which says that

$$p_*(c|\Sigma(f)) = tr_B^E(c).$$

Now suppose that there is no fiberwise oriented GMF on E. Then we need to stabilize by taking the fiberwise product with a large even dimensional sphere

$$S^{2N} = D^{2N}_- \cup D^{2N}_+.$$

By the framed function theorem there is a fiberwise framed function

$$f : E \times D^{2N}_- \to (-1, 0]$$

taking the boundary to 0. By definition, the higher FR–torsion of $C_*(f)$ is equal to the higher torsion of E:

$$\tau_{2k}(E) = \tau_{2k}(C_*(f)).$$

Taking $-f$ on the upper hemisphere we get a fiberwise oriented GMF

$$f \cup -f : E \times S^{2N} \to \mathbb{R}.$$

By the Splitting Lemma and the Suspension Theorem, the higher torsion of $E \times S^{2N}$ is equal to twice the higher torsion of E:

$$\tau_{2k}(E \times S^{2N}) = \tau_{2k}(E \times D^{2N}_-) + \tau_{2k}(E \times D^{2N}_+, \partial) = (1 + (-1)^{2N})\tau_{2k}(E).$$

But the Framing Principle tells us that:

$$\tau_{2k}(E \times S^{2N}) = \tau_{2k}(C_*(f \cup -f)) + (-1)^k\zeta(2k+1)p_*\left(\tfrac{1}{2}ch_{2k}(\gamma_{-f} \otimes \mathbb{C})\right).$$

The first term is zero by the Splitting Lemma and involution property:

$$\tau_{2k}(C_*(f \cup -f)) = \tau_{2k}(C_*(f)) + \tau_{2k}(C_*(-f)) = 0.$$

In the second term, γ_{-f} is isomorphic to the restriction to $\Sigma(-f)$ of the vertical tangent bundle of $E^* = E \times D_+^{2N}$. Consequently,

$$p_* \left(\tfrac{1}{2} ch_{2k}(\gamma_{-f} \otimes \mathbb{C})\right) = tr_B^{E^*} \left(\tfrac{1}{2} ch_{2k}(T^v E^* \otimes \mathbb{C})\right) = \frac{1}{(2k)!} T_{2k}(E)$$

by (6.4). Now divide both sides by 2.

$$\tau_{2k}(E) = \frac{1}{2}\tau_{2k}(E \times S^{2N}) = \frac{1}{2}(-1)^k \zeta(2k+1) \frac{1}{(2k)!} T_{2k}(E).$$

□

Now suppose that the fiber M is even dimensional and oriented with boundary.

THEOREM 6.7 (torsion for M^{2n} oriented with boundary). *Suppose that M^{2n} is an oriented even dimensional manifold and $M \to E \to B$ is a smooth bundle where the action of $\pi_1 B$ on the rational homology of both M and ∂M is upper triangular. Suppose also that $\pi_1 B$ preserves the orientation of M. Then the higher FR–torsion invariants $\tau_{2k}(E) \in H^{4k}(B; \mathbb{R})$ are defined and given by*

$$\tau_{2k}(E) = \underbrace{\frac{1}{2}\tau_{2k}(\partial E)}_{exotic\ component} + \underbrace{\frac{1}{2}(-1)^k \zeta(2k+1) \frac{1}{(2k)!} T_{2k}(E)}_{tangential\ component}.$$

REMARK 6.8 (exotic and tangential torsion). The two terms in the above formula can be viewed as the *exotic* and *tangential* components of the higher FR–torsion. The tangential component is a tangential homotopy invariant. The exotic component always comes from a bundle with odd dimensional closed fiber.

PROOF. Suppose first that there is a fiberwise oriented GMF $f : E \to \mathbb{R}$. Then the Framing Principle for f and $-f$ gives us:

$$\tau_{2k}(E) = \tau_{2k}(C_*(f)) + (-1)^k \zeta(2k+1) p_* \left(\tfrac{1}{2} ch_{2k}(\gamma_f \otimes \mathbb{C})\right)$$

$$\tau_{2k}(E, \partial E) = \tau_{2k}(C_*(-f)) + (-1)^k \zeta(2k+1) p_* \left(\tfrac{1}{2} ch_{2k}(\gamma_{-f} \otimes \mathbb{C})\right).$$

By the Splitting Lemma we always have:

$$\tau_{2k}(E) = \tau_{2k}(E, \partial E) + \tau_{2k}(\partial E).$$

Adding these three equations we get:

$$2\tau_{2k}(E) = \tau_{2k}(\partial E) + (-1)^k \zeta(2k+1) p_* \left(\tfrac{1}{2} ch_{2k}(T^v E \otimes \mathbb{C})\right)$$

and we are done. (See the proof of Theorem 6.6.)

If the function f does not exist we need to take the product with a high even dimensional sphere S^{2N}. This double all three terms in our theorem. □

6.3. Unoriented fibers

If the fiber M is unoriented we can pass to an oriented disk bundle over E. Here we take an oriented 1–disk bundle.

Every unoriented manifold M has a 2–fold oriented covering $\widetilde{M}$ whose rational homology is equal to the sum

$$H_*(\widetilde{M}; \mathbb{Q}) \cong H_*(M; \mathbb{Q}) \oplus H_*(M; \mathbb{Q}^-)$$

where $\mathbb{Q}^-$ is a suitable twisted coefficient system on M. This formula also holds with $\mathbb{Q}$ replaced by $\mathbb{R}$ or $\mathbb{C}$. If M is closed, the covering $\widetilde{M}$ bounds a 1–dimensional disk bundle over M which we call JM. Applying this to each fiber of a smooth bundle $E \to B$ we get a 2–fold covering $\widetilde{E}$ of E which forms the boundary of a disk bundle JE.

LEMMA 6.9 (comparing E and JE). a) $T_{2k}(E) = T_{2k}(JE)$.
b) $\tau_{2k}(E) = \tau_{2k}(JE)$ *if it is defined.*
c) *If* $\tau_{2k}(E)$ *and* $\tau_{2k}(E;\mathbb{C}^-)$ *are both defined then*

$$\tau_{2k}(\widetilde{E}) = \tau_{2k}(E) + \tau_{2k}(E;\mathbb{C}^-).$$

PROOF. (a) follows from the fact that the Pontrjagin classes of a real line bundle are all zero. (b) is the positive suspension theorem (3.2). (c) is a special case of the transfer formula Theorem 3.30 since $\mathrm{Ind}_1^{Z_2}\mathbb{C} = \mathbb{C} \oplus \mathbb{C}^-$. □

Suppose that M is odd dimensional and closed. Then JE is a smooth bundle with oriented even dimensional fibers. Consequently, its higher torsion is given by Theorem 6.7. Using Lemma 6.9 above, this calculation can be expressed as follows.

THEOREM 6.10 (torsion for M^{2n+1} closed, unoriented). *If* $\tau_{2k}(E;\mathbb{C})$ *and* $\tau_{2k}(E;\mathbb{C}^-)$ *are both defined then*

$$\tau_{2k}(E;\mathbb{C}^\pm) = \frac{1}{2}\tau_{2k}(\widetilde{E}) \pm \frac{1}{2}(-1)^k\zeta(2k+1)\frac{1}{(2k)!}T_{2k}(E).$$

If M is even dimensional, closed and unoriented, the best we can do is the following.

THEOREM 6.11 (torsion for M^{2n} closed, unoriented).

$$\tau_{2k}(E) + \tau_{2k}(E;\mathbb{C}^-) = (-1)^k\zeta(2k+1)\frac{1}{(2k)!}T_{2k}(E).$$

PROOF. By Lemma 6.9(c) and the torsion for oriented even dimensional fibers (Theorem 6.6) we have:

$$\tau_{2k}(E) + \tau_{2k}(E;\mathbb{C}^-) = \tau_{2k}(\widetilde{E}) = \frac{1}{2}(-1)^k\zeta(2k+1)\frac{1}{(2k)!}T_{2k}(\widetilde{E}),$$

where

$$T_{2k}(\widetilde{E}) = 2T_{2k}(JE) = 2T_{2k}(E)$$

by Proposition 6.5 applied to JE and Lemma 6.9(a). □

If M has a boundary then $\widetilde{E}$ is still a 2–fold covering of E but it forms only part of the boundary of JE. Let J^*E be the disk bundle JE with the corners rounded. Then the boundary of J^*E is an oriented bundle over B

$$\partial J^*E = \widetilde{E} \cup J\partial E$$

with oriented closed manifold fibers $\partial J^*M = \widetilde{M} \cup J\partial M$. As before, we have

$$\tau_{2k}(E) = \tau_{2k}(J^*E)$$

if the terms are defined. If M is odd dimensional then J^*E is oriented with even dimensional fibers.

THEOREM 6.12 (torsion for M^{2n+1} unoriented with boundary). *If the fiber M of $p : E \to B$ is odd dimensional then*

$$\tau_{2k}(E) = \frac{1}{2}\tau_{2k}(\widetilde{E} \cup J\partial E) + \frac{1}{2}(-1)^k\zeta(2k+1)\frac{1}{(2k)!}T_{2k}(J^*E)$$

when all terms are defined.

For the sake of completeness, we make the following trivial observation.

PROPOSITION 6.13 (torsion for M^{2n} unoriented). If M is even dimensional and unoriented then

$$\tau_{2k}(E) = \tau_{2k}(JE)$$

assuming the terms are defined.

REMARK 6.14. In all cases, the calculation of the higher torsion is reduced to the case of an oriented odd dimensional fiber (plus the tangential homotopy invariant T_{2k}).

6.4. Vertical normal disk bundle

We need to generalize this to a higher dimensional disk bundles. Since E is compact, we can always find a fiberwise embedding of E into $B \times \mathbb{R}^{2N}$ for N sufficiently large. Choose such an embedding and let $D(E)$ be the unit normal disk bundle of E in $B \times \mathbb{R}^{2N}$. Let $S(E)$ be the unit normal sphere bundle. We call $D(E), S(E)$ the *vertical normal disk bundle* and the *vertical normal sphere bundle*, respectively.

The vertical tangent bundle of $D(E)$ is trivial by construction. Consequently,

$$T_{2k}(D(E)) = 0.$$

Since $D(E)$ has even dimensional fibers, Theorem 6.7 gives us the following.

THEOREM 6.15 (torsion of $D(E)$ for M closed, oriented). *Suppose that the fiber M is a closed oriented manifold and the action of $\pi_1 B$ on the rational homology of M is upper triangular. Then*

$$\tau_{2k}(E) = \tau_{2k}(D(E)) = \frac{1}{2}\tau_{2k}(S(E)).$$

If M has a boundary then

$$\partial D(E) = D(\partial E) \cup S(E)$$

and these two subsets meet at an angle along the set

$$D(\partial E) \cap S(E) = S(\partial E)$$

whose vertical tangent bundle is stably parallelizable. Since the fibers of $S(\partial E)$ are closed and even dimensional its higher torsion is trivial if it is defined.

THEOREM 6.16 (torsion of $D(E)$ for M oriented with boundary). *Suppose that M is oriented with boundary and the action of $\pi_1 B$ on the rational homology of both M and ∂M is upper triangular. Then*

$$\tau_{2k}(E) = \frac{1}{2}\tau_{2k}(\partial E) + \frac{1}{2}\tau_{2k}(S(E)).$$

PROOF. Let $D^*(E)$ be $D(E)$ with the corners rounded. Then

(1) $\tau_*(E) = \tau_*(D^*(E))$ since $D(E)$ is a linear disk bundle over E (3.2).

(2) $\tau_*(D^*(E)) = \frac{1}{2}\tau_*(\partial D^*(E))$ since $D^*(E)$ has even dimensional parallelizable fibers (6.7).
(3) Since $\partial D^*(E) = D(E)|\partial E \cup S(E)$, the gluing formula (3.17) gives

$$\tau_*(\partial D^*(E)) = \tau_*(D(E)|\partial E) + \tau_*(S(E)) - \tau_*(S(E)|\partial E).$$

(4) $\tau_*(D(E)|\partial E) = \tau_*(\partial E)$, again by (3.2).
(5) $\tau_*(S(E)|\partial E) = 0$ since $S(E)|\partial E$ is a bundle with stably parallelizable closed even dimensional fibers (6.6).

□

If we compare this with Theorem 6.7 we get:

COROLLARY 6.17 (torsion of $S(E)$ for M^{2n} oriented). Suppose that M is oriented and even dimensional and that the action of $\pi_1 B$ on the rational homology of both M and ∂M is upper triangular. Then

$$\tau_{2k}(S(E)) = (-1)^k \zeta(2k+1)\frac{1}{(2k)!}T_{2k}(E).$$

REMARK 6.18 (generalizing the transfer theorem). Examination of this corollary reveals that it is a transfer formula. It says that the higher torsion of $S(E)$ over B is the transfer of the higher torsion of $S(E)$ over E. However, it is not an example of the transfer theorem since the fibers of $S(E)$ over E (odd dimensional spheres) are not acyclic. This suggests that the transfer theorem may hold whenever the euler characteristic of the fiber is zero. (It does not hold otherwise.)

CHAPTER 7

The Stability Theorem

The purpose of this chapter is to clarify the statement of the stability theorem for diffeomorphisms as requested by the referee. These statements are taken from [**Igu88**] and [**Igu02**] with additional explanations and proofs.

7.1. Definitions

For manifolds with boundary we consider the space

$$Diff_\partial(M) := Diff(M \text{ rel } \partial M)$$

of diffeomorphisms of M which are the identity on ∂M. To stabilize we pass to $Diff_\partial(M \times I^n)$ for large n in a way that we will describe.

The stability theorem for diffeomorphisms comes from the stability theorem for the *concordance space* of M which is defined to be

$$\mathcal{C}(M) := Diff(M \times I \text{ rel } M \times 0 \cup \partial M \times I).$$

In this construction we speak of M as being "horizontal" and I as being "vertical." Thus, concordances of M are diffeomorphisms of $M \times I$ which are the identity on the bottom and sides. Restriction to the top ($M \times 1$) gives the *restriction map*

$$r : \mathcal{C}(M) \to Diff_\partial(M)$$

which is a fibration over its image. The appendix of [**Cer68**] gives a good explanation of this fact. The fiber of the restriction map is $Diff_\partial(M \times I)$.

There is a "suspension map"

$$\sigma_+ : \mathcal{C}(M) \to \mathcal{C}(M \times I)$$

which has the property that it is "laterally symmetric" in the sense that $s \circ \sigma_+ = \sigma_+$ where s is the involution on $(M \times I) \times I$ given by $s(x, s, t) = (x, 1 - s, t)$.

Iteration of the suspension map σ_+ gives the *stable concordance space*

$$\mathcal{P}(M) := \varinjlim \mathcal{C}(M \times I^n).$$

7.2. Stability for $\mathcal{C}(M)$

THEOREM 7.1 (Stability Theorem). [**Igu88**] *The positive suspension map*

$$\sigma_+ : \mathcal{C}(M) \to \mathcal{C}(M \times I)$$

k-connected if

$$\dim M \geq \max(2k + 7, 3k + 4). \tag{7.1}$$

REMARK 7.2 (stable range). This means that the map

$$\pi_k \mathcal{C}(M) \to \pi_k \mathcal{C}(M \times I) \cong \pi_k \mathcal{P}(M)$$

induced by σ_+ is an isomorphism in the *stable range* given by

$$\dim M \geq \max(2k+9, 3k+7). \tag{7.2}$$

and an epimorphism in the *surjective stability range* given by (7.1) above.

7.3. Involution

The fibration sequence

$$Diff_\partial(M \times I) \xrightarrow{j} \mathcal{C}(M) \xrightarrow{r} Diff_\partial(M) \tag{7.3}$$

is equivariant with respect to the involution t given by

(1) *reflection* or *lateral involution* on $Diff_\partial(M \times I)$: $t(g) = s \circ g \circ s$ where $s \in Diff(M \times I)$ is given by $s(x,t) = (x, 1-t)$,
(2) *vertical involution* on $\mathcal{C}(M)$: $t(g) = (r(g)^{-1} \times id_I) \circ s \circ g \circ s$ and
(3) *inversion* on $Diff_\partial(M)$: $t(g) = g^{-1}$.

Rationally, the homotopy groups decompose into positive and negative eigenspaces.

PROPOSITION 7.3 (positive part of $\pi_k Diff$). The inclusion map $Diff_\partial(M \times I) \hookrightarrow \mathcal{C}(M)$ induces an isomorphism

$$\pi_k^+ Diff_\partial(M \times I) \otimes \mathbb{Q} \cong \pi_k^+ \mathcal{C}(M) \otimes \mathbb{Q}.$$

PROOF. This follows from the positive part of the long exact homotopy sequence of (7.3) since the involution on (7.3) acts as inversion (-1) on $\pi_k Diff_\partial(M)$ for all $k \geq 0$. □

THEOREM 7.4 (quasi-stability of $\pi_k^\pm \mathcal{C}(M)$). *In the surjective stability range (7.1) we have*

$$\pi_k^+ \mathcal{C}(M) \otimes \mathbb{Q} \cong \pi_k^- \mathcal{C}(M \times I) \otimes \mathbb{Q} \tag{7.4}$$

In the stable range (7.2) we have

$$\pi_k^- \mathcal{C}(M) \otimes \mathbb{Q} \cong \pi_k^+ \mathcal{C}(M \times I) \otimes \mathbb{Q} \tag{7.5}$$

REMARK 7.5 (stability of $\pi_k^\pm \mathcal{C}(M)$). We must increase the dimension of M in (7.4) in order for $\pi_k^- \mathcal{C}(M \times I) \otimes \mathbb{Q}$ to be stable. Thus $\pi_k^+ \mathcal{C}(M) \otimes \mathbb{Q}$ is stable (and $\pi_k^+ Diff_\partial(M \times I)$ is stable) if

$$\dim(M) \geq \max(2k+8, 3k+6) \tag{7.6}$$

and $\pi_k^- \mathcal{C}(M) \otimes \mathbb{Q}$ is stable in the stable range $\dim M \geq \max(2k+9, 3k+7)$.

PROOF. Since suspension anticommutes with the involution, both mappings are isomorphisms in the stable range and epimorphisms in the surjective stability range. However the map (7.4) is always a monomorphism since the composition

$$Diff_\partial(M \times I) \xrightarrow{j} \mathcal{C}(M) \xrightarrow{\sigma_+} \mathcal{C}(M \times I) \xrightarrow{r} Diff_\partial(M \times I) \tag{7.7}$$

is equal to $1 + t$ (t = reflection) and therefore an isomorphism on $\pi_k^+ \otimes \mathbb{Q}$ for all k. □

Since the decomposition of $\pi_k \mathcal{C}(M)$ into positive and negative parts depends on the parity of $\dim M$ we take even dimensional manifolds to define the decomposition of $\pi_k \mathcal{P}(M)$.

DEFINITION 7.6 (decomposition of $\pi_k\mathcal{P}(M)$).

$$\pi_k^{\pm}\mathcal{P}(M)\otimes\mathbb{Q}:=\pi_k^{\pm}\mathcal{C}(M\times I^N)$$

where $N=4k+10-\dim M$ (or any other sufficiently large number with the same parity as $\dim M$).

COROLLARY 7.7 (desuspension of $\pi_k^-\mathit{Diff}$). In the surjective stability range $\dim(M\times I)\geq\max(2k+8,3k+5)$ we have

$$\pi_k^-\mathit{Diff}_\partial(M\times I)\otimes\mathbb{Q}\cong\pi_0^-\mathit{Diff}_\partial(M\times I^{k+1})\otimes\mathbb{Q}.$$

PROOF. Look at the long exact homotopy sequence of the fibration

$$\mathit{Diff}_\partial(M\times I^2)\xrightarrow{j}\mathcal{C}(M\times I)\xrightarrow{r}\mathit{Diff}_\partial(M\times I)\tag{7.8}$$

This is equivariant with respect to reflection/vertical involution/inversion as in (7.3). But it is also equivariant with respect to reflection on all three terms because of the extra I factor. We use this lateral involution in this proof.

The positive part of $\pi_k\mathit{Diff}_\partial(M\times I^2)$ goes to the right and the positive part of $\pi_k\mathit{Diff}_\partial(M\times I)$ goes to the left by (7.7). In the surjective stability range this accounts for all of $\pi_k\mathcal{C}(M\times I)$ since the image of the suspension map σ_+ is reflection invariant. So, we must have

$$\pi_k^-\mathit{Diff}_\partial(M\times I)\otimes\mathbb{Q}\cong\pi_{k-1}^-\mathit{Diff}_\partial(M\times I^2)\otimes\mathbb{Q}.$$

The corollary follows by induction. □

7.4. Disks and spheres

We consider the special case of D^n and S^n.

COROLLARY 7.8 (vanishing of $\pi_k^-\mathit{Diff}_\partial(D^n)$).

$$\pi_k^-\mathit{Diff}_\partial(D^n)\otimes\mathbb{Q}=0$$

when $n\geq\max(2k+8,3k+5)$.

PROOF. Applying Corollary 7.7 in the case $M=D^{n-1}$ we get

$$\pi_k^-\mathit{Diff}_\partial(D^n)\otimes\mathbb{Q}=\pi_0\mathit{Diff}_\partial(D^{n+k})\otimes\mathbb{Q}.$$

But this is zero since

$$\pi_0\mathit{Diff}_\partial(D^{n+k})\cong\Gamma_{n+k+1}\tag{7.9}$$

is the group of exotic $n+k+1$–spheres which is a finite group by [**KM63**], [**Cer68**]. □

To identify the positive part of $\pi_k\mathit{Diff}_\partial(D^n)$ we need the lemma:

LEMMA 7.9 (calculation of $\pi_k^{\pm}\mathcal{P}(*)$).

$$\pi_k^{\epsilon}\mathcal{P}(*)\otimes\mathbb{Q}\cong\begin{cases}\mathbb{Q} & \text{if } \epsilon=+ \text{ and } k\equiv 3 \mod 4\\ 0 & \text{otherwise}\end{cases}$$

REMARK 7.10 (Hatcher's construction). Hatcher constructed a mapping

$$G/O\to B\mathcal{P}(*)$$

which was proved by Bökstedt [**Bök84**] to be a rational homotopy equivalence. Bökstedt actually proved that G/O is rationally equivalent to $\Omega Wh^{diff}(*)$ and quoted Waldhausen's theorem [**Wal87**] that $\Omega Wh^{diff}(X)\simeq B\mathcal{P}(X)$. (See [**Igu02**] for the details of this construction.)

THEOREM 7.11 (calculation of $\pi_k^+ Diff_\partial(D^n)$). *Suppose $n \geq \max(2k+9, 3k+7)$. Then*

$$\pi_k^+ Diff_\partial(D^n) \otimes \mathbb{Q} \cong \begin{cases} \mathbb{Q} & \text{if } n \text{ is odd and } k \equiv 3 \mod 4 \\ 0 & \text{otherwise} \end{cases}$$

PROOF. In the stable range ($n \geq \max(2k+9, 3k+7)$ by Remark 7.5) we have

$$\pi_k^+ Diff_\partial(D^n) \cong \pi^\epsilon \mathcal{P}(*)$$

where $\epsilon = (-1)^{n-1}$. (Use $M = D^{n-1}$ in Definition 7.6.) Rationally this is given by Lemma 7.9. □

COROLLARY 7.12 (rational cohomology of $BDiff_\partial(D^n)$). For $k \leq \min\left(\frac{n-5}{2}, \frac{n+1}{3}\right)$, $BDiff_\partial(D^n)$ has the rational k–homotopy type of BO for n odd and the rational k–homotopy type of a point for n even. Consequently, through degree $\left\lfloor \min\left(\frac{n-7}{2}, \frac{n-2}{3}\right)\right\rfloor$ we have an isomorphism

$$H^*(BDiff_\partial(D^n); \mathbb{Q}) \cong \begin{cases} H^*(BO; \mathbb{Q}) & \text{if } n \text{ is odd} \\ 0 & \text{if } n \text{ is even} \end{cases}$$

PROOF. Let $SDiff_\partial(D^n)$ be the identity component of $Diff_\partial(D^n)$. Then, for n odd, we have rational k–equivalences

$$BO \leftarrow G/O \rightarrow B\mathcal{P}(*) \leftarrow B\mathcal{C}(D^n) \rightarrow BSDiff_\partial(D^n) \rightarrow BDiff_\partial(D^n)$$

The four spaces in the middle are simply connected using Cerf's theorem [**Cer70**] that $\mathcal{C}(D^n)$ is connected for $n \geq 5$. Thus, they have isomorphic rational cohomology through degree $k-1$. This extends to the left since $BO \simeq BSO \times \mathbb{R}P^\infty$. To the right we use the fact that $\Gamma_{n+1} = \pi_0 Diff_\partial(D^n)$ is a finite abelian group as noted in (7.9) above. Furthermore, the action of Γ_{n+1} on $BSDiff_\partial(D^n)$ is homopically trivial since it corresponds to conjugation in $Diff_\partial(D^n)$ which is homopy commutative. This proves the corollary for odd n. The even case is easy. □

The passage from disks to spheres is easy because of the following formula.

$$Diff(S^n) \simeq O(n+1) \times Diff_\partial(D^n).$$

Furthermore, this homotopy equivalence gives a homotopy equivalence of classifying spaces since it is induced by a homomorphism $O(n+1) \times G \rightarrow Diff(S^n)$ where $G \simeq Diff_\partial(D^n)$ is the subgroup of diffeomorphisms which fix a neighborhood of ∂D^n.

COROLLARY 7.13 (rational cohomology of $BDiff(S^n)$). For $k \leq \min\left(\frac{n-5}{2}, \frac{n+1}{3}\right)$, $BDiff(S^n)$ has the rational k–homotopy type of $BO \times BO$ for n odd and BO for n even. Consequently, through degree $\left\lfloor \min\left(\frac{n-7}{2}, \frac{n-2}{3}\right)\right\rfloor$, we have an isomorphism

$$H^*(BDiff(S^n); \mathbb{Q}) \cong \begin{cases} H^*(BO; \mathbb{Q}) \otimes H^*(BO; \mathbb{Q}) & \text{if } n \text{ is odd} \\ H^*(BO; \mathbb{Q}) & \text{if } n \text{ is even} \end{cases}$$

7.5. Relation to higher torsion

Strictly speaking, the higher FR–torsion invariants are only defined on the two–fold covering $BDiff_0(S^n)$ of $BDiff(S^n)$ where $Diff_0(S^n)$ is the group of orientation preserving diffeomorphisms of S^n. However,

$$Diff_0(S^n) \simeq SO(n+1) \times Diff_\partial(D^n).$$

So, in the stable range, $BDiff_0(S^n)$ has the same rational (and real) cohomology as $BDiff(S^n)$. Thus, the higher FR–torsion classes

$$\tau_{2k} \in H^{4k}(BDiff(S^n); \mathbb{R})$$

come from unique real cohomology classes on $BDiff(S^n)$ in the stable range

$$4k \leq \min\left(\frac{n-7}{2}, \frac{n-2}{3}\right).$$

We finally come to our updated version of the Hsiang-Farrell Theorem mentioned in the introduction.

THEOREM 7.14 (higher torsion as generators of $H^*(BDiff(S^n))$. *If n is even then the real cohomology of $BDiff(S^n)$ is a polynomial algebra in the higher FR–torsion classes τ_{2k} in the stable range given above.*

If n is odd then, in the stable range given above, $H^*(BDiff(S^n); \mathbb{R})$ is a polynomial algebra in the higher FR–torsion classes τ_{2k} and the Pontrjagin classes p_j coming from the projection $BDiff(S^n) \to BO(n+1)$.

PROOF. If we restrict the higher torsion classes $\tau_2, \tau_4, \cdots, \tau_{2k}$ to $BO(n+1)$ they generate the same subring of the real cohomology of $BO(n+1)$ as the Pontrjagin classes $p_1, p_2, \cdots, p_k$ according to Theorem 4.1 and Remark 4.2. This together with Corollary 7.13 prove the theorem for n even.

If n is odd, we also need to use the fact [**Igu02**] that the pullbacks of the higher torsion classes τ_{2k} under Hatcher's construction are proportional to the Chern character and thus freely generate the real cohomology of G/O. □

REMARK 7.15 (question). The topological decomposition of the classifying space $BDiff(S^{2n+1})$ given in Corollary 7.13 does not correspond to the decomposition of its cohomology given by the Pontrjagin classes and the higher FR–torsion classes since these higher torsion classes are nontrivial on both the linear part $BO(2n+2)$ and the tangentially trivial part $BDiff_\partial(D^{2n+1})$.

As characteristic classes of smooth sphere bundles, the Pontrjagin classes measure the tangential homotopy type of the bundle. They are trivial on $BDiff_\partial(D^{2n+1})$. The question arises: What geometric property characterizes smooth sphere bundles with trivial higher torsion? What is the subspace of $BDiff(S^{2n+1})$ on which the higher torsion classes all vanish?

Note that these question can be phrased in terms of higher analytic torsion since Goette [**Goe**] has shown that the higher (nonequivariant) analytic torsion is proportional to the higher FR–torsion for smooth odd dimensional sphere bundles.

Bibliography

[BG75] J.C. Becker and D.H. Gottlieb, *The transfer and fiber bundles*, Topology **14** (1975), 1–12.

[BG01] Jean-Michel Bismut and Sebastian Goette, *Families torsion and Morse functions*, Astérisque (2001), no. 275, x+293.

[BL95] Jean-Michel Bismut and John Lott, *Flat vector bundles, direct images and higher real analytic torsion*, J. Amer. Math. Soc. **8** (1995), no. 2, 291–363.

[Bök84] M. Bökstedt, *The rational homotopy type of $\Omega Wh^{Diff}(*)$*, LNM 1051, Springer-Verlag, 1984, pp. 25–37.

[Bor74] A. Borel, *Stable real cohomology of arithmetic groups*, Ann. Sci. ENS **7** (1974), 235–272.

[Bro59] Edgar H. Brown, Jr., *Twisted tensor products. I*, Ann. of Math. (2) **69** (1959), 223–246.

[BT82] R. Bott and L. Tu, *Differential forms in algebraic topology*, Graduate Texts in Math., vol. 82, Springer-Verlag, New York, 1982.

[Bun] U. Bunke, *Higher analytic torsion of sphere bundles and continuous cohomology of $Diff(S^{2n-1})$*, preprint, math.DG/9802100.

[Car81] Jeffrey L. Caruso, *A simpler approximation to QX*, Trans. Amer. Math. Soc. **265** (1981), no. 1, 163–167.

[Cer68] Jean Cerf, *Sur les difféomorphismes de la sphère de dimension trois* ($\Gamma_4 = 0$), Springer-Verlag, Berlin, 1968.

[Cer70] ———, *La stratification naturelle des espaces de fonctions différentiables réelles et le théorème de la pseudo-isotopie*, Inst. Hautes Études Sci. Publ. Math. (1970), no. 39, 5–173.

[Dup76] J. Dupont, *Simplicial deRham cohomology and characteristic classes of flat bundles*, Topology **15** (1976), 233–245.

[DWW03] W. Dwyer, M. Weiss, and B. Williams, *A parametrized index theorem for the algebraic K-theory Euler class*, Acta Math. **190** (2003), no. 1, 1–104.

[FH78] F. T. Farrell and W. C. Hsiang, *On the rational homotopy groups of the diffeomorphism groups of discs, spheres and aspherical manifolds*, Algebraic and Geometric Topology (Proc. Sympos. Pure Math., Stanford Univ., Stanford, Calif., 1976), Part 1, Amer. Math. Soc., Providence, R.I., 1978, pp. 325–337.

[Goe] Sebastian Goette, *Morse theory and higher torsion invariants I*, math.DG/0111222.

[HW73] Allen Hatcher and John Wagoner, *Pseudo-isotopies of compact manifolds*, Société Mathématique de France, Paris, 1973, With English and French prefaces, Astérisque, No. 6.

[Igu] Kiyoshi Igusa, *Twisted cochains and higher torsion*, math.AT/0212383.

[Igu87] ———, *The space of framed functions*, Trans. Amer. Math. Soc. **301** (1987), no. 2, 431–477.

[Igu88] ———, *The stability theorem for smooth pseudoisotopies*, K-Theory **2** (1988), no. 1–2, vi+355.

[Igu90] ———, *C^1 local parametrized Morse theory*, Topology Appl. **36** (1990), no. 3, 209–231.

[Igu02] ———, *Higher Franz-Reidemeister Torsion*, AMS/IP Studies in Advance Mathematics, vol. 31, International Press, 2002.

[IK93] Kiyoshi Igusa and John Klein, *The Borel regulator map on pictures. II. An example from Morse theory*, K-Theory **7** (1993), no. 3, 225–267.

[IK95] ———, *Higher Franz-Reidemeister torsion. I. The algebra of filtered chain complexes*, preprint, 1995.

[Kad80] T.V. Kadeishvilli, *On the homology theory of fiber spaces*, Uspekhi Mat.Nauk **35** (1980), 183–188, Translated in Russ Math. Surv. **35** (1980), 231–238.

[Kle89] John Klein, *The cell complex construction and higher R-torsion for bundles with framed Morse function*, Ph.D. thesis, Brandeis University, 1989.

[Kle93] ———, *Higher Franz-Reidemeister torsion and the Torelli group*, Mapping Class Groups and Moduli Spaces, Contemp. Math., vol. 150, American Math. Soc., 1993, pp. 195–204.

[KM63] Michel A. Kervaire and John W. Milnor, *Groups of homotopy spheres. I*, Ann. of Math. (2) **77** (1963), 504–537.

[KT74] F. Kamber and P. Tondeur, *Characteristic invariants of foliated bundles*, Manuscripta Math. **11** (1974), 51–89.

[Ma97] Xiaonan Ma, *Formes de torsion analytique et familles de submersions*, C. R. Acad. Sci. Paris Sér. I Math. **324** (1997), no. 2, 205–210.

[Mil83] John Milnor, *On polylogarithms, Hurwitz zeta functions, and the Kubert identities*, Enseign. Math. (2) **29** (1983), no. 3-4, 281–322.

[Mil86] Edward Y. Miller, *The homology of the mapping class group*, J. Differential Geom. **24** (1986), no. 1, 1–14.

[Mor87] Shigeyuki Morita, *Characteristic classes of surface bundles*, Invent. Math. **90** (1987), no. 3, 551–577.

[Mor01] ———, *Geometry of Characteristic Classes*, Translations of Mathematical Monographs, vol. 199, American Math. Soc., 2001.

[Mum83] David Mumford, *Towards an enumerative geometry of the moduli space of curves*, Arithmetic and geometry, Vol. II, Birkhäuser Boston, Boston, MA, 1983, pp. 271–328.

[Qui85] Daniel Quillen, *Superconnections and the Chern character*, Topology **24** (1985), 89–95.

[Wal85] Friedhelm Waldhausen, *Algebraic K-theory of spaces*, Algebraic and Geometric Topology (New Brunswick, N.J., 1983), Lecture Notes in Math. 1126, Springer, Berlin, 1985, pp. 318–419.

[Wal87] ———, *Algebraic K-theory of spaces, concordance and stable homotopy theory*, Algebraic Topology and Algebraic K-theory (William Browder, ed.), Annals of Math. Studies, vol. 113, 1987, pp. 392–417.

Editorial Information

To be published in the *Memoirs*, a paper must be correct, new, nontrivial, and significant. Further, it must be well written and of interest to a substantial number of mathematicians. Piecemeal results, such as an inconclusive step toward an unproved major theorem or a minor variation on a known result, are in general not acceptable for publication. Papers appearing in *Memoirs* are generally at least 80 and not more than 200 published pages in length. Papers less than 80 or more than 200 published pages require the approval of the Managing Editor of the Transactions/Memoirs Editorial Board.

As of May 31, 2005, the backlog for this journal was approximately 11 volumes. This estimate is the result of dividing the number of manuscripts for this journal in the Providence office that have not yet gone to the printer on the above date by the average number of monographs per volume over the previous twelve months, reduced by the number of volumes published in four months (the time necessary for preparing a volume for the printer). (There are 6 volumes per year, each containing at least 4 numbers.)

A Consent to Publish and Copyright Agreement is required before a paper will be published in the *Memoirs*. After a paper is accepted for publication, the Providence office will send a Consent to Publish and Copyright Agreement to all authors of the paper. By submitting a paper to the *Memoirs*, authors certify that the results have not been submitted to nor are they under consideration for publication by another journal, conference proceedings, or similar publication.

Information for Authors

Memoirs are printed from camera copy fully prepared by the author. This means that the finished book will look exactly like the copy submitted.

The paper must contain a *descriptive title* and an *abstract* that summarizes the article in language suitable for workers in the general field (algebra, analysis, etc.). The *descriptive title* should be short, but informative; useless or vague phrases such as "some remarks about" or "concerning" should be avoided. The *abstract* should be at least one complete sentence, and at most 300 words. Included with the footnotes to the paper should be the 2000 *Mathematics Subject Classification* representing the primary and secondary subjects of the article. The classifications are accessible from `www.ams.org/msc/`. The list of classifications is also available in print starting with the 1999 annual index of *Mathematical Reviews*. The Mathematics Subject Classification footnote may be followed by a list of *key words and phrases* describing the subject matter of the article and taken from it. Journal abbreviations used in bibliographies are listed in the latest *Mathematical Reviews* annual index. The series abbreviations are also accessible from `www.ams.org/publications/`. To help in preparing and verifying references, the AMS offers MR Lookup, a Reference Tool for Linking, at `www.ams.org/mrlookup/`. When the manuscript is submitted, authors should supply the editor with electronic addresses if available. These will be printed after the postal address at the end of the article.

Electronically prepared manuscripts. The AMS encourages electronically prepared manuscripts, with a strong preference for $\mathcal{A}_{\mathcal{M}}\mathcal{S}$-LaTeX. To this end, the Society has prepared $\mathcal{A}_{\mathcal{M}}\mathcal{S}$-LaTeX author packages for each AMS publication. Author packages include instructions for preparing electronic manuscripts, the *AMS Author Handbook*, samples, and a style file that generates the particular design specifications of that publication series. Though $\mathcal{A}_{\mathcal{M}}\mathcal{S}$-LaTeX is the highly preferred format of TeX, author packages are also available in $\mathcal{A}_{\mathcal{M}}\mathcal{S}$-TeX.

Authors may retrieve an author package from e-MATH starting from `www.ams.org/tex/` or via FTP to `ftp.ams.org` (login as `anonymous`, enter username as password, and type `cd pub/author-info`). The *AMS Author Handbook* and the *Instruction Manual* are available in PDF format following the author packages link from `www.ams.org/tex/`. The author package can be obtained free of charge by sending email to pub@ams.org (Internet) or from the Publication Division, American Mathematical Society, 201 Charles St., Providence, RI 02904, USA. When requesting an author package, please specify $\mathcal{A}_\mathcal{M}\mathcal{S}$-LaTeX or $\mathcal{A}_\mathcal{M}\mathcal{S}$-TeX, Macintosh or IBM (3.5) format, and the publication in which your paper will appear. Please be sure to include your complete mailing address.

Sending electronic files. After acceptance, the source file(s) should be sent to the Providence office (this includes any TeX source file, any graphics files, and the DVI or PostScript file).

Before sending the source file, be sure you have proofread your paper carefully. The files you send must be the EXACT files used to generate the proof copy that was accepted for publication. For all publications, authors are required to send a printed copy of their paper, which exactly matches the copy approved for publication, along with any graphics that will appear in the paper.

TeX files may be submitted by email, FTP, or on diskette. The DVI file(s) and PostScript files should be submitted only by FTP or on diskette unless they are encoded properly to submit through email. (DVI files are binary and PostScript files tend to be very large.)

Electronically prepared manuscripts can be sent via email to pub-submit@ams.org (Internet). The subject line of the message should include the publication code to identify it as a Memoir. TeX source files, DVI files, and PostScript files can be transferred over the Internet by FTP to the Internet node e-math.ams.org (130.44.1.100).

Electronic graphics. Comprehensive instructions on preparing graphics are available at www.ams.org/jourhtml/graphics.html. A few of the major requirements are given here.

Submit files for graphics as EPS (Encapsulated PostScript) files. This includes graphics originated via a graphics application as well as scanned photographs or other computer-generated images. If this is not possible, TIFF files are acceptable as long as they can be opened in Adobe Photoshop or Illustrator. No matter what method was used to produce the graphic, it is necessary to provide a paper copy to the AMS.

Authors using graphics packages for the creation of electronic art should also avoid the use of any lines thinner than 0.5 points in width. Many graphics packages allow the user to specify a "hairline" for a very thin line. Hairlines often look acceptable when proofed on a typical laser printer. However, when produced on a high-resolution laser imagesetter, hairlines become nearly invisible and will be lost entirely in the final printing process.

Screens should be set to values between 15% and 85%. Screens which fall outside of this range are too light or too dark to print correctly. Variations of screens within a graphic should be no less than 10%.

Inquiries. Any inquiries concerning a paper that has been accepted for publication should be sent directly to the Electronic Prepress Department, American Mathematical Society, 201 Charles St., Providence, RI 02904, USA.

Editors

Titles in This Series

836 **H. G. Dales and A. T.-M. Lau,** The second duals of Beurling algebras, 2005

835 **Kiyoshi Igusa,** Higher complex torsion and the framing principle, 2005

834 **Ken'ichi Ohshika,** Kleinian groups which are limits of geometrically finite groups, 2005

833 **Greg Hjorth and Alexander S. Kechris,** Rigidity theorems for actions of product groups and countable Borel equivalence relations, 2005

832 **Lee Klingler and Lawrence S. Levy,** Representation type of commutative Noetherian rings III: Global wildness and tameness, 2005

831 **K. R. Goodearl and F. Wehrung,** The complete dimension theory of partially ordered systems with equivalence and orthogonality, 2005

830 **Jason Fulman, Peter M. Neumann, and Cheryl E. Praeger,** A generating function approach to the enumeration of matrices in classical groups over finite fields, 2005

829 **S. G. Bobkov and B. Zegarlinski,** Entropy bounds and isoperimetry, 2005

828 **Joel Berman and Paweł M. Idziak,** Generative complexity in algebra, 2005

827 **Trevor A. Welsh,** Fermionic expressions for minimal model Virasoro characters, 2005

826 **Guy Métivier and Kevin Zumbrun,** Large viscous boundary layers for noncharacteristic nonlinear hyperbolic problems, 2005

825 **Yaozhong Hu,** Integral transformations and anticipative calculus for fractional Brownian motions, 2005

824 **Luen-Chau Li and Serge Parmentier,** On dynamical Poisson groupoids I, 2005

823 **Claus Mokler,** An analogue of a reductive algebraic monoid whose unit group is a Kac-Moody group, 2005

822 **Stefano Pigola, Marco Rigoli, and Alberto G. Setti,** Maximum principles on Riemannian manifolds and applications, 2005

821 **Nicole Bopp and Hubert Rubenthaler,** Local zeta functions attached to the minimal spherical series for a class of symmetric spaces, 2005

820 **Vadim A. Kaimanovich and Mikhail Lyubich,** Conformal and harmonic measures on laminations associated with rational maps, 2005

819 **F. Andreatta and E. Z. Goren,** Hilbert modular forms: Mod p and p-adic aspects, 2005

818 **Tom De Medts,** An algebraic structure for Moufang quadrangles, 2005

817 **Javier Fernández de Bobadilla,** Moduli spaces of polynomials in two variables, 2005

816 **Francis Clarke,** Necessary conditions in dynamic optimization, 2005

815 **Martin Bendersky and Donald M. Davis,** V_1-periodic homotopy groups of $SO(n)$, 2004

814 **Johannes Huebschmann,** Kähler spaces, nilpotent orbits, and singular reduction, 2004

813 **Jeff Groah and Blake Temple,** Shock-wave solutions of the Einstein equations with perfect fluid sources: Existence and consistency by a locally inertial Glimm scheme, 2004

812 **Richard D. Canary and Darryl McCullough,** Homotopy equivalences of 3-manifolds and deformation theory of Kleinian groups, 2004

811 **Ottmar Loos and Erhard Neher,** Locally finite root systems, 2004

810 **W. N. Everitt and L. Markus,** Infinite dimensional complex symplectic spaces, 2004

809 **J. T. Cox, D. A. Dawson, and A. Greven,** Mutually catalytic super branching random walks: Large finite systems and renormalization analysis, 2004

808 **Hagen Meltzer,** Exceptional vector bundles, tilting sheaves and tilting complexes for weighted projective lines, 2004